Climate Change Management

Series editor

Walter Leal Filho, Faculty of Life Sciences, Research and Transfer Centre,
Hamburg University of Applied Sciences, Hamburg, Germany

More information about this series at http://www.springer.com/series/8740

Colleen Murphy · Paolo Gardoni
Robert McKim
Editors

Climate Change and Its Impacts

Risks and Inequalities

Editors
Colleen Murphy
College of Law
University of Illinois at Urbana-Champaign
Champaign, IL
USA

Robert McKim
Department of Religion
University of Illinois at Urbana-Champaign
Urbana, IL
USA

Paolo Gardoni
Department of Civil and Environmental
 Engineering
University of Illinois at Urbana-Champaign
Urbana, IL
USA

ISSN 1610-2002 ISSN 1610-2010 (electronic)
Climate Change Management
ISBN 978-3-319-77543-2 ISBN 978-3-319-77544-9 (eBook)
https://doi.org/10.1007/978-3-319-77544-9

Library of Congress Control Number: 2018938643

© Springer International Publishing AG, part of Springer Nature 2018
This work is subject to copyright. All rights are reserved by the Publisher, whether the whole or part of the material is concerned, specifically the rights of translation, reprinting, reuse of illustrations, recitation, broadcasting, reproduction on microfilms or in any other physical way, and transmission or information storage and retrieval, electronic adaptation, computer software, or by similar or dissimilar methodology now known or hereafter developed.
The use of general descriptive names, registered names, trademarks, service marks, etc. in this publication does not imply, even in the absence of a specific statement, that such names are exempt from the relevant protective laws and regulations and therefore free for general use.
The publisher, the authors and the editors are safe to assume that the advice and information in this book are believed to be true and accurate at the date of publication. Neither the publisher nor the authors or the editors give a warranty, express or implied, with respect to the material contained herein or for any errors or omissions that may have been made. The publisher remains neutral with regard to jurisdictional claims in published maps and institutional affiliations.

Printed on acid-free paper

This Springer imprint is published by the registered company Springer International Publishing AG part of Springer Nature
The registered company address is: Gewerbestrasse 11, 6330 Cham, Switzerland

Contents

Part I Introduction

1 **Risks and Values: New and Interconnected Challenges of Climate Change** 3
   Colleen Murphy, Paolo Gardoni and Robert McKim

Part II The Paris Agreement, Policy and Climate Justice

2 **Climate Change in the 21st Century: Looking Beyond the Paris Agreement** 15
   Donald J. Wuebbles

3 **Cumulative Harm as a Function of Carbon Emissions** 39
   John Nolt

4 **Justice in Mitigation After Paris** 53
   Darrel Moellendorf

5 **Utilitarianism, Prioritarianism, and Climate Change: A Brief Introduction** 69
   Matthew D. Adler

Part III Natural Hazards, Resilience and Mitigation

6 **Assessing Climate Change Impacts on Hurricane Hazards** 93
   David V. Rosowsky

7 **Climate Change, Heavy Precipitation and Flood Risk in the Western United States** 109
   Eric P. Salathé Jr. and Guillaume Mauger

8 **The Impact of Climate Change on Resilience of Communities Vulnerable to Riverine Flooding** 129
   Xianwu Xue, Naiyu Wang, Bruce R. Ellingwood and Ke Zhang

9   **Planning for Community Resilience Under Climate Uncertainty** ... 145
    Ross B. Corotis

Part IV   Responding to Climate Change: Mitigation and Adaptation

10 **Climate Change Governance and Local Democracy: Synergy or Dissonance** 163
    Emmanuel O. Nuesiri

11 **Sea Level Rise and Social Justice: The Social Construction of Climate Change Driven Migrations** 181
    Elizabeth Marino

12 **Recovery After Disasters: How Adaptation to Climate Change Will Occur** 195
    Robert B. Olshansky

Part V   Responding to Climate Change: Priorities, Perspectives, and Solutions

13 **The Climate-Change Challenge to Human-Drawn Boundaries** 211
    Eric T. Freyfogle

14 **Neoliberal (Mis)Management of Earth-Time and the Ethics of Climate Justice** 233
    Michael S. Northcott

15 **Human Capital in a Climate-Changed World** 251
    Shi-Ling Hsu

16 **A Wild Solution for Climate Change** 269
    Thomas E. Lovejoy

Part I
Introduction

Chapter 1
Risks and Values: New and Interconnected Challenges of Climate Change

Colleen Murphy, Paolo Gardoni and Robert McKim

Abstract This introductory chapter provides a brief summary of the main aims of the book. We also provide an overview of the structure of the volume as a whole and the main points of each chapter.

Introduction

Climate change is one of the most important and pressing contemporary global challenges for the international community. Climate change is modifying the likelihood and magnitude of natural hazards around the world and creating new vulnerabilities (Gardoni et al. 2016). These hazards include heat waves and their effects on wildfires and droughts; severe precipitation and its effects on floods and large snowfall events; and hurricanes. Climate change is also causing sea level rise that affects coastal communities where large and vulnerable populations often reside. It is estimated that $70–$100 billion will be needed by developing countries to adapt to the anticipated impacts of climate change. There is a clear need for a deeper understanding of the consequences of climate change, of the attendant natural hazards, and of their social impact.

Topics to which particular attention is paid in this book include:

1. Scientific understanding of the effects of climate change on the likelihood and magnitude of natural hazards;

C. Murphy (✉) · P. Gardoni · R. McKim
University of Illinois at Urbana-Champaign, College of Law, 504 East Pennsylvania Ave., Champaign, IL 61820, USA
e-mail: colleenm@illinois.edu

P. Gardoni
e-mail: gardoni@illinois.edu

R. McKim
e-mail: rmckim@illinois.edu

© Springer International Publishing AG, part of Springer Nature 2018
C. Murphy et al. (eds.), *Climate Change and Its Impacts*,
Climate Change Management, https://doi.org/10.1007/978-3-319-77544-9_1

2. Determination of the contribution that each person alive today is making to climate change;
3. Selection of the kind of ethical framework and lines of reasoning needed to evaluate behavior that contributes to climate change;
4. Assessment of civil infrastructure vulnerabilities as they are exacerbated by climate change as well as probabilistic predictions and stochastic formulations of intensified extreme load demands on infrastructure;
5. Development of new design criteria, codes and standards that can be put in place to help mitigate the impacts of climate change;
6. Identification of inequalities in vulnerability among communities and discussion of how these are exacerbated or diminished by climate change;
7. Resilience assessment for coastal communities exposed to hurricanes, storm surges and coastal floods affected by climate change;
8. Policies that can be put in place to help mitigate the impacts of climate change;
9. Cultural shifts and reevaluation of our priorities that might help humanity to respond adequately to climate change.

The basic premise of this book is that an appropriate and comprehensive response to climate change requires the technical expertise of engineers and scientists; the legal, cultural, political, environmental and economic expertise of social scientists and legal scholars; and the moral expertise of ethicists, including philosophers. In keeping with this premise, the book brings climate scientists, engineers, and urban planners into conversation with legal scholars, geographers, anthropologists and ethicists. The chapters provide a broad overview of how climate change is conceptualized by academics in all of these fields.

Structure and Overview of the Book

The book is organized into five parts. Part I consists in this introduction. Part II begins with an up to date account of what science is telling us about climate change and its consequences. Then the focus shifts to the moral implications of climate change in general and of the Paris Agreement in particular. The Paris Agreement entered into force on November 4, 2016, and to date 167 countries have ratified it. It "requires all Parties to put forward their best efforts through 'nationally determined contributions' (NDCs) and to strengthen these efforts in the years ahead. This includes requirements that all Parties report regularly on their emissions and on their implementation efforts" (United Nations 2015). Part III turns to a specific set of risks that are and will continue to be affected by climate change: risks from natural hazards. Part IV begins our discussion of responses to climate change, probing the issues of mitigation and adaptation. Also included here are analyses of how, where, and by whom mitigation as well as adaptation efforts should be made and prescriptions for ways to approach climate adaptation and mitigation. In Part V additional responses are considered. These include new perspectives on how to understand the problem and possible partial solutions to the problem, as well as some reflection about what motivates people to respond appropriately.

Part II: The Paris Agreement, Policy and Climate Justice

In Chap. 2, "Climate change in the 21st century: looking beyond the Paris Agreement," atmospheric scientist Donald J. Wuebbles provides a detailed and up to date summary of (a) the many lines of evidence that have led to the scientific consensus that the planet is warming due to human activity and especially because of the accumulation of CO_2 in the atmosphere, (b) the consequences of climate change such as sea-level rise and the increased frequency, intensity, and duration of severe weather events, (c) some projected consequences for the 21st century and beyond, and (d) an outline of the main options that humanity faces.

How should we think about the ways in which the behavior of individuals contributes to climate change? In Chap. 3, "Cumulative harm as a function of carbon emissions," philosopher John Nolt takes up the issue of the quantification of the harm associated with climate change. Establishing causal responsibility informs judgments of moral responsibility, but is not sufficient for moral responsibility (e.g., knowledge may matter too). Nolt focuses on the impact of cumulative emissions. He starts from the premise that global average temperature during any given future time period will increase directly and continuously with our cumulative emissions from the present through that period. Harm is also, he argues, going to increase directly and continuously with average global temperature increases. So harm increases directly and continuously with our cumulative emissions. What this means is that even small increases in emissions may cause significant harm.

The Paris Agreement and climate change more generally raise foundational questions of policy and of ethics. In Chap. 4, "Justice in mitigation after Paris," philosopher Darrel Moellendorf invokes a distinction between international justice, which deals with how burdens are distributed among states, and intergenerational justice, which deals with how burdens are distributed across generations. Because the Paris Agreement involved a decentralized approach in which states decided what climate change mitigation steps to undertake, pledges made by states can be assumed to be consistent with their own development and poverty eradication objectives. Consequently, adherence to the Paris Agreement probably is consistent with the requirements of international justice, at least in the short term. In the long term, however, an intergenerational collective action problem looms. It arises from the increasingly ambitious pledges that the Paris Agreement requires. If the cost of renewable energy does not fall rapidly enough, or if it is not understood to do so by the public, international cooperation is threatened. Moellendorf considers how this collective action problem is best understood and mechanisms to solve it.

Intergenerational justice and international justice are not the only two lens that may be used to evaluate the ethics of the Paris Agreement and policies enacted to fulfill it. In Chap. 5, "Prioritarianism and climate change," legal scholar Matthew Adler presents a comparative analysis of prioritarianism and utilitarianism as frameworks for evaluating climate change mitigation policies. Prioritarianism gives special consideration to the impact of policies on the well-being of the worse-off and does not engage in discounting. Utilitarianism considers the impact of a policy on well-being overall, discounting impacts that are farther into the future. Utilitarianism remains the prevailing framework for climate economics. However, Adler argues that prioritarianism is a more ethically defensible framework although more work needs to be done in modeling from a prioritiarian perspective.

Part III: Natural Hazards, Resilience and Mitigation

As Wuebbles' chapter (Chap. 2) explains, the consequences of climate change are many. These include exacerbating natural hazards such as hurricanes and flooding. As the global community prepares for climate change and its consequences, it is important to determine how these hazards will be affected by climate change and how best to mitigate, and be resilient in the face of, such hazards.

Chapters 6, 7, and 8 focus on the impact of climate change on particular hazards. In Chap. 6, "Assessing climate change impacts on hurricane hazards," civil engineer and risk analyst David V. Rosowsky addresses the question of whether the predicted climate change scenarios will have a tangible effect on the hurricane hazard. The chapter conclusively shows that the worst-case scenario in climate change will have a clear effect on the hurricane hazard on the US coastline. The results from this chapter can be used by decision and policy makers as well as insurers/re-insurers and risk portfolio managers. They can also be used to develop optimal mitigation strategies that make best use of resources and properly balance the risks faced by communities.

In Chap. 7, "Climate change, heavy precipitation and flood risk in the western United States," climate scientists Eric Salathé and Guillaume Mauger examine the role of climate change in flood risk. Instead of developing predictions and risk assessments based on historical data, the occurrence of climate change requires deriving such predictions and assessments from climate models and downscaling methods. Downscaling methods are used to obtain local flood predictions that are needed for community risk and resilience analysis. The chapter discusses both statistical and dynamical downscaling and their implications for flood predictions. The chapter ends by presenting a case study that shows the impact on a flood plain of sea level rise, reduced snowpack and higher intensity precipitation extremes.

In the United States the flooding of rivers has long been the cause of significant damage to the built and natural environments and has resulted in much social harm. In Chap. 8, "The impact of climate change on resilience of communities vulnerable to riverine flooding," an interdisciplinary team consisting of civil engineers, risk analysts and an atmospheric scientist considers riverine flooding, which is caused by a river exceeding its capacity due to excessive rainfall or significant snow melt over a short period of time. Xianwu Xue, Naiyu Wang, Bruce R. Ellingwood, and Ke Zhang develop a new modeling framework for flood hazard analysis that incorporates the effects of climate change. This framework uses a hydrological model within a hydraulic analysis. The hydrological model is used to simulate hydrological processes at a course spatial resolution. The hydraulic analysis is used to compute flood variables (like localized flood depths, velocities and inundated areas) considering a finer spatial resolution. The new framework is calibrated and validated using the Wolf River Basin in Shelby County, Tennessee.

In Chap. 9, "Planning for community resilience under climate uncertainty," civil engineer and risk analyst Ross B. Corotis challenges the probabilistic models used in risk analysis of future hazards. The premise of his challenge is that such probabilistic models have been traditionally calibrated using historical data. However, the changes brought by climate change in the likelihood of occurrence and magnitude of the stressors to a community call for a reevaluation of such models. The chapter also notes the importance of considering communities as a whole in contrast to single structures considered in isolation. To promote community resilience, the chapter puts forward the concept of adaptive management and defines the participatory methods by which community mitigation actions can be developed.

Part IV: Responding to Climate Change: Mitigation and Adaptation

Fulfilling the aspirations of the Paris Agreement requires countries to take specific actions. Part IV includes chapters that consider the policies and strategies that are being, will be, and should be adopted to aid communities at all scales from the local to the national and beyond in both mitigating and adapting to the consequences of climate change.

In Chap. 10, "Climate change governance and local democracy: synergy or dissonance," geographer Emmanuel Nuesiri focuses on the question of local governance in climate change mitigation and adaptation programs and policies. He looks specifically at programs targeting emissions reductions stemming from efforts to prevent deforestation, to encourage reforestation and sustainable management of forests specifically in developing countries, known as the (REDD+) initiative. Priority in such programs should be given to local democratic participation as a way of ensuring that REDD+ programs benefit local people. However, Nuesiri offers a cautionary tale of the UN-REDD funded Nigeria-REDD program. He highlights its failure to ensure robust local democratic participation and its insufficient engagement with local government authorities. The chapter ends by recommending that UN-REDD programs not only interact with NGOs but also with local authorities.

In Chap. 11, "Sea level rise and social justice: the ethics of climate change driven migration," anthropologist Elizabeth Marino draws on case studies from the United States to illustrate the way existing social policies and colonial legacies influence who is vulnerable to displacement from climate change. She discusses how the social, political and legal context shape which natural events become disasters. Marino outlines the criteria used to determine who counts as a climate refugee, the legal and political consequences that follow from being excluded or included, and the environmental techniques for protecting communities from rising sea levels. She then focuses on the decisions that will shape which individuals and communities living in coastal areas will be displaced from rising sea levels.

Next the discussion moves from displacement to adaptation with a focus on the case of sea level rise. In Chap. 12, "Recovery after disasters: how adaptation to climate change will occur," urban planner Robert B. Olshansky argues that, in most cases, communities will notice changes in the sea level on the occasion of particular events such as coastal storms and storm surges rather than on account of a continuous background increase in the sea level. The fact that sea level rise will come to people's attention in this way will shape the adaptation process, which will be part of the long-term post-disaster recovery. Given this feature of the adaptation process, this chapter describes the phases and players in the post-disaster recovery along with its challenges and the disruptions it will bring. In keeping with themes raised by both Nuesiri and Marino, Olshansky argues that a successful recovery requires involvement of the affected citizens.

Part V: Responding to Climate Change: Priorities, Perspectives, and Solutions

The final set of chapters introduces some new perspectives on how to understand climate change and how to respond to it, some reflection about what motivates people to respond, and some proposals about steps that would contribute to finding a solution.

In Chap. 13, "The climate-change challenge to human-drawn boundaries," legal scholar Eric T. Freyfogle proposes that the best way to approach climate change and its consequences is in terms of a comprehensive goal of ensuring that the landscapes around us are healthy, diverse, and resilient, while also facilitating human flourishing. Pursuing this goal requires the modification of core elements of our culture. Instead of an emphasis on the rights of individuals, including individual landowners, we ought to focus on the common good of the land community. And we should rethink the institution of private property, conceiving of it as having the purpose of promoting the welfare of the land community. These changes in turn require that we rethink human-drawn boundaries at all levels. In some respects boundaries are less relevant today. After all, the consequences of climate change do not conform to our boundaries. But in other respects boundaries are more relevant: in particular, establishing and preserving healthy, diverse, and resilient landscapes may sometimes require local control and territorial autonomy.

How are people moved by the harms that climate change is generating and will generate? In Chap. 14, "Neoliberal (mis)management of Earth-time and the ethics of climate justice," moral theologian Michael S. Northcott argues that the main reasons that people of faith are concerned about climate change include their compassion for the vulnerable, their concern about people who are already being harmed by climate change, and their concern for their own children and grandchildren. These are among the findings derived from interviews with congregation members in Scottish churches that have a record of promoting ecological responsibility. The interviewees were accordingly less impressed by a neo-liberal emphasis on what is economically most attractive or by what course of action a utilitarian summing up of costs and benefits would dictate or by short-term performance targets.

In Chap. 15, "Human capital in a climate-changed world," legal scholar Shi-Ling Hsu examines the issue of economic development in an era of climate change. Hsu argues that it is a mistake to see increased fossil fuel use as necessary to development. He contends that economic development can proceed in conjunction with efforts to remove fossil fuel subsidies. Moreover, resources that would otherwise not be available due to such subsidies can instead be used to focus on what will be necessary to maintain development in the midst of climate change: education. This shift would have the extra benefit of compensating those most likely to be harmed by climate change.

Finally, in Chap. 16, "A wild solution for climate change," conservation biologist Thomas E. Lovejoy begins by providing an up to date account of the consequences of climate change for biodiversity. He summarizes various changes on land and in the oceans that are already occurring. These include flowering plants blossoming earlier and earlier animal migrations. He outlines what the best research leads us to expect given the likely effects of climate change on habitat, especially when this is combined with heavy human use of landscapes. We can expect dislocations and extinctions and unpleasant surprises when poorly understood thresholds are crossed. Lovejoy considers solutions including the obvious one of moving away as quickly as possible from fossil fuels. At the end of his paper he mentions some research that supports the "wild solution" mentioned in the title of his chapter. The key idea is that ecosystem restoration might pull enough CO_2 out of the atmosphere to combat climate change.

Closing Reflections

Responding adequately to climate change requires a collective effort of society. It requires integrating and synthesizing the scientific understanding of climate change, models to predict the impact of climate change on the natural and built environments, an understanding of the implications of climate change for individuals' well-being and the way vulnerability shapes these implications, the formulation of public policies, and the existence of political will. As the chapters in this volume illustrate, interdisciplinary research and discussions among the different stakeholders should aim to develop successful strategies that are technically sound and that promote international justice, intergenerational justice and environmental justice.

Chapters in this volume point to the areas where further research is needed for our collective success in responding to climate change. Adaptation and mitigation are the dominant strategies for responding to climate change. It remains to be seen where the limits of each strategy lie, and whether the source of such limits is technical in nature or social or both (Adger et al. 2009). It also remains to be seen whether there will be the political will to respond to the crisis of climate change, and if there are ideas and principles that might move and inspire people and that have not yet been articulated. Furthermore, it remains to be seen whether political institutions will be able to prioritize these problems that are global in character and intergenerational in temporal scope (Gardiner 2011).

Finally, climate change raises important questions of trade-offs and potential moral conflicts. Laudato Si', Pope Francis' impressive recent encyclical on the environment, asks us to hear both "the cry of the poor" and "the cry of the earth" (Francis 2015, Sect. 49). On the one hand, it seems that we can simultaneously respond to both cries. The poor are among the most vulnerable to climate change; hence steps to address climate change that emphasize the most vulnerable will at a minimum be compatible with contributing to solving both problems at once (Thomas and Twyman 2005). On the other hand, steps to ameliorate the problems of the poor may be bad for the earth. As more poor people become better off there is characteristically more consumption and increased greenhouse gas emissions. Those of us who are better off can consume less, giving others a chance to make their way out of poverty without making things worse in terms of total emissions and total human impact. But a failure on the part of people who have options may force everyone into a tragic situation in which efforts to combat inequality will continue to exacerbate climate change. It remains to be seen whether a way forward that does justice to both of these fundamentally important concerns—the plight of the poor and climate change—will be found.

Acknowledgements This book builds upon the presentations, discussions, and related outcomes from the workshop *Climate Change and Its Impact: Risks and Inequalities* that took place at the University of Illinois at Urbana-Champaign on March 10–11, 2016. This workshop was made possible through grants from the Department of Civil and Environmental Engineering (CEE)—Research Thrust Development Program, the College of Liberal Arts and Sciences Conference

Support Program, and the Illinois International Conference Grant Program. The Women and Gender in Global Perspectives Program (WGGP), the College of Law and the MAE Center: Creating a Multi-hazard Approach to Engineering provided additional support. The opinions and findings presented in this book are those of the authors of each chapter and do not necessarily reflect the views of the sponsors or the editors.

References

Adger WN, Dessai S, Goulden M, Hulme M, Lorenzoni I, Nelson DR, Naess LO, Wolf J, Wreford A (2009) Are there social limits to adaptation to climate change? Clim Change 93(3–4): 335–354

Gardiner S (2011) A perfect moral storm: the ethical tragedy of climate change. Oxford University Press, New York

Gardoni P, Murphy C, Rowell A (2016) Risk analysis of natural hazards: interdisciplinary challenges and integrated solutions. Springer, Dordrecht

Pope Francis (2015) Laudato si: on care for our common home. http://w2.vatican.va/content/francesco/en/encyclicals/documents/papa-francesco_20150524_enciclica-laudato-si.html

Thomas DSG, Twyman C (2005) Equity and justice in climate change adaptation amongst natural-resource-dependent societies. Glob Environ Change 15(2):115–124

United Nations (2015) United Nations framework convention on climate change, http://unfccc.int/paris_agreement/items/9485.php

Part II
The Paris Agreement, Policy and Climate Justice

Chapter 2
Climate Change in the 21st Century: Looking Beyond the Paris Agreement

Donald J. Wuebbles

Abstract The science is clear that the Earth's climate, including that of the United States, is changing, changing much more rapidly than generally occurs naturally, and it is happening primarily because of human activities. This chapter discusses the science underlying climate change and the current understanding of how our planet is being affected. In addition to the global analysis, there is special attention given to the findings for the United States. Humanity is already feeling the effects from increasing intensity of certain types of extreme weather and from sea level rise that are fueled by the changing climate. Climate change affects many sectors of our society, including threats on human health and well-being. Climate change will, absent other factors, amplify some of the existing threats we now face. The effects on humanity are already significant, costing us many billions of dollars each year along with the effects on human lives and health. Policy to respond to climate change is imperative—we have three choices, mitigation, adaptation, or suffering. Right now we are doing some of all three. The Paris Agreement begins the process internationally of really doing something to slow down change. But the current agreement is just the beginning and we will need to do much more.

Introduction

The science is clear: the Earth's climate is changing, it is changing extremely rapidly, and the evidence shows it is happening primarily because of human activities (IPCC 2013, 2014; Melillo et al. 2014; UKRS-NAS 2014; and the thousands of papers referenced in these assessments). Climate change is happening now—it is not just a problem for the future—and it is happening throughout the world. There are many indicators of the changing climate. Surface temperature is just one of them. Trends in the severity of certain types of severe weather events are

---

D. J. Wuebbles (✉)
The Harry E. Preble Professor of Atmospheric Sciences, Department of Atmospheric Sciences, University of Illinois, 105 S. Gregory St., Urbana, IL 61801, USA
e-mail: Wuebbles@illinois.edu

© Springer International Publishing AG, part of Springer Nature 2018
C. Murphy et al. (eds.), *Climate Change and Its Impacts*,
Climate Change Management, https://doi.org/10.1007/978-3-319-77544-9_2

increasing. Sea levels are also rising because of the warming oceans and because of the melting land ice. Observations show that the climate is changing extremely rapidly, about ten times more rapidly than natural changes in climate based on paleoclimatic observations of the changes that occurred since the end of the last ice age. And the evidence clearly points to climate changes over the last half century as being primarily due to human activities, especially the burning of fossil fuels and also land use change, especially through deforestation. As a result, it is not surprising that many national and world leaders have concluded that climate change, often referred to as global warming in the media, has become one of the most important issues facing humanity.

There is essentially no debate in the peer-reviewed scientific literature (or in the national and international assessments of the science prepared by hundreds of scientists) about the large changes occurring in the Earth's climate and the fact that these changes are occurring as a response to human activities. Natural factors such as changes in the energy output of the Sun have always affected our climate in the past and continue to do so today; but over the last century, human activities have become the dominant influence in producing many, if not most, of the observed changes occurring in our current climate.

People throughout the world are already feeling the effects from increasing intensity of certain types of extreme weather and from sea level rise that are fueled by the changing climate. Prolonged periods of heat and heavy downpours, and in some regions, floods and in others, drought, are affecting our health, agriculture, water resources, energy and transportation infrastructure, and much more.

The harsh reality is that the present amount of climate change is already dangerous and will become far more dangerous in the coming decades. Climate change is itself likely to increase the risks for impacts on human society and on ecosystems, and the more intense extreme events associated with a changing climate pose a serious risk to human health.

The chapter begins with a discussion of the changes happening and projected to happen in the climate system and a summary of the underlying scientific basis for the human cause for these changes. Much more on each of these topics, and the projections of future changes in climate, can be found in the international (IPCC 2013, 2014) and U.S. National Climate (Melillo et al. 2014) assessments of the science mentioned earlier. The connections between potential impacts and the changing climate are then examined, with a special focus on the United States based on discussion in the 3rd National Climate Assessment (NCA: Melillo et al. 2014). Issues associated with mitigation and adaptation policy, including the effects of the Paris Agreement are then assessed.

Our Changing Climate

The fifth assessment report (AR5) of the Intergovernmental Panel on Climate Change (IPCC 2013, 2014) is the most comprehensive analysis to date of the science of climate change and how it is affecting our planet. Over 800 scientists and other experts were involved in the four volumes of this assessment. Similarly, the 3rd U.S. National Climate Assessment (Melillo et al. 2014) is the most comprehensive analysis to date of how climate change is affecting the United States now and how it could affect it in the future. A team of more than 300 scientists and other experts (see complete list online at http://nca2014.globalchange.gov), guided by a 60-member National Climate Assessment and Development Advisory Committee, produced the assessment. Stakeholders involved in the development of the assessment included decision-makers from the public and private sectors, resource and environmental managers, researchers, representatives from businesses and non-governmental organizations, and the general public. The resulting report went through extensive peer and public review before publication, including two sets of reviews by the National Academy of Sciences. The NCA collects, integrates, and assesses observations and research from around the country, helping us to see what is actually happening and understand what it means for our lives, our livelihoods, and our future. The report includes analyses of impacts on seven sectors—human health, water, energy, transportation, agriculture, forests, and ecosystems—and the interactions among sectors at the national level. The report also assesses key impacts on all parts of the United States and evaluated for specific regions: Northeast, Southeast and Caribbean, Midwest, Great Plains, Southwest, Northwest, Alaska, Hawaii and Pacific Islands, as well as the country's coastal areas, oceans, and marine resources. By being so comprehensive, the NCA aim is to help inform Americans' choices and decisions about investments, where to build and where to live, how to create safer communities and secure our own and our children's future. The 4th National Climate Assessment is now underway and will be published in 2018.

Climate is defined as long-term averages and variations in weather measured over multiple decades. The Earth's climate system includes the land surface, atmosphere, oceans, and ice. Scientists from around the world have compiled the evidence that the climate is changing, changing much more rapidly than tends to occur naturally (by a factor of ten or more relative to the natural changes that occurred following the end of the last ice age 20,000 years ago), and that it is changing because of human activities; these conclusions are based on observations from satellites, weather balloons, thermometers at surface stations, ice cores, and many other types of observing systems that monitor the Earth's weather and climate. A wide variety of independent observations give a consistent picture of a warming world. There are many indicators of this change, not just atmospheric surface temperature. For example, ocean temperatures are also rising, sea level is rising, Arctic sea ice is decreasing, most glaciers are decreasing, Greenland and Antarctic land ice is decreasing, and atmospheric humidity is increasing.

Climate Change Effects on Temperature

Temperatures at the surface, in the troposphere [the active weather layer extending from the ground to about 8–16 km (5–10 miles altitude)], and in the oceans have all increased over recent decades. Consistent with our scientific understanding, the largest increases in temperature are occurring closer to the poles, especially in the Arctic (this is especially related to ice-albedo feedback, which, as snow and ice decrease, indicates that the exposed surface will absorb more solar radiation rather than reflect it back to space). Snow and ice cover have decreased in most areas on Earth. Atmospheric water vapor (H_2O) is increasing in the lower atmosphere, because a warmer atmosphere can hold more water (the basic physics is captured by the Clausius–Clapeyron equation, which provides the relationship between temperature and available water vapor). Sea levels are also increasing. All of these findings are based on observations.

As seen in Fig. 1, global annual average temperature (as measured over both land and oceans) has increased by more than 0.8 °C (1.5 °F) since 1880 (through 2012). Since then, 2014 was the warmest year on record, but this was greatly eclipsed by 2015, when a strong El Niño event (unusually warm water in the eastern portion of the Pacific Ocean) added to the effects of climate change. So far, it looks like 2016 will be warmer still. While there is a clear long-term global warming trend, some years do not show a temperature increase relative to the previous year, and some years show greater changes than others. These year-to-year fluctuations in temperature are related to natural processes, such as the effects of ocean events like El Niños and La Niñas, and the cooling effects of atmospheric emissions from volcanic eruptions. At the local to regional scale, changes in climate can be influenced by natural variability for a few decades (Deser et al. 2012). Globally, natural variations can be as large as human-induced climate change over

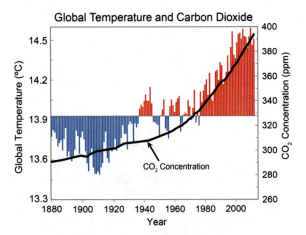

Fig. 1 Changes in observed globally-averaged temperature since 1880. Red bars show temperatures above the long-term average, and blue bars indicate temperatures below the long-term average. The black line shows the changes in atmospheric carbon dioxide (CO_2) concentration in parts per million (ppm) over the same time period (Melillo et al. 2014; temperature data from NOAA National Climate Data Center)

timescales of up to a decade (Karl et al. 2015). However, changes in climate at the global scale observed over the past 50 years are far larger than can be accounted for by natural variability (IPCC 2013).

While there has been widespread warming over the past century, not every region has warmed at the same pace (Fig. 2). A few regions, such as the North Atlantic Ocean and some parts of the U.S. Southeast, have even experienced cooling over the last century as a whole, though the U.S. Southeast has warmed over recent decades. This is due to the stronger influence of internal variability over smaller geographic regions and shorter time scales. Warming during the first half of the last century occurred mostly in the Northern Hemisphere. The last three decades have seen greater warming in response to accelerating increases in heat-trapping gas concentrations, particularly at high northern latitudes, and over land as compared to the oceans. These findings are not surprising given the larger heat capacity of the oceans leading to land-ocean differences in warming and the ice-albedo feedback

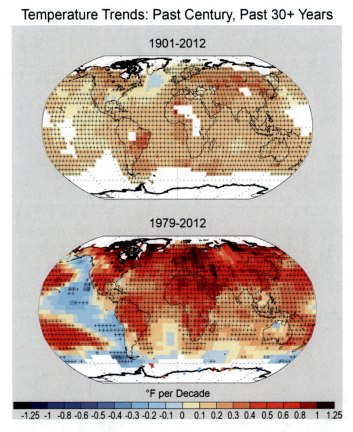

Fig. 2 Surface temperature trends for the period 1901–2012 (top) and 1979–2012 (bottom) from NOAA National Climate Data Center's surface temperature product. Updated from Vose et al. (2012). From Melillo et al. (2014)

leading to larger warming at higher latitudes. As a result, land areas can respond to the changes in climate much more rapidly than the ocean areas even though the forcing driving a change in climate occurs equally over land and the oceans.

Even if the surface temperature had never been measured, scientists could still conclude with high confidence that the global temperature has been increasing because multiple lines of evidence all support this conclusion. Figure 3 shows a number of examples of the indicators that show the climate on Earth is changing very rapidly over the last century. Temperatures in the lower atmosphere and oceans have increased, as have sea level and near-surface humidity. Basic physics tells us that a warmer atmosphere can hold more water vapor; this is exactly what is measured from the satellite data showing that humidity is increasing. Arctic sea ice,

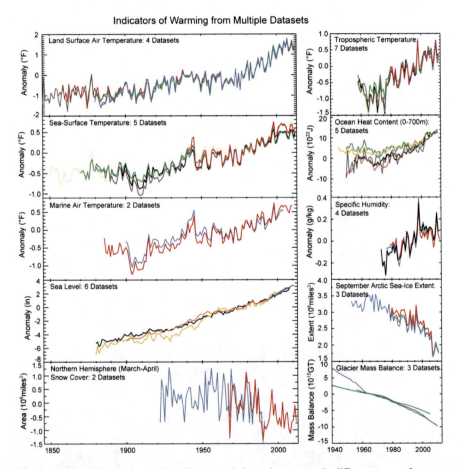

Fig. 3 Observed changes, as analyzed by many independent groups in different ways, of a range of climate indicators. All of these are in fact changing as expected in a warming world. Further details underpinning this diagram can be found at http://www.ncdc.noaa.gov/bams-state-of-the-climate/. From Melillo et al. (2014)

mountain glaciers, and Northern Hemisphere spring snow cover have all decreased. Over 90% of the glaciers in the world are decreasing at very significant rates. The amount of ice on the largest masses of ice on our planet, on Greenland and Antarctica, are decreasing. As with temperature, many scientists and associated research groups have analyzed each of these indicators and come to the same conclusion: all of these changes paint a consistent and compelling picture of a warming planet.

Climate Change Effects on Precipitation

Precipitation is perhaps the most societally relevant aspect of the hydrological cycle and has been observed over global land areas for over a century. However, spatial scales of precipitation are small (e.g., it can rain several inches in Washington, DC, but not a drop in nearby Baltimore) and this makes interpretation of the point-measurements difficult. Based upon a range of efforts to create global averages, there does not appear to have been significant changes in globally averaged precipitation since 1900 (although as we will discuss later there has been a significant trend for an increase in precipitation coming as larger events). However, in looking at total precipitation there are strong geographic trends including a likely increase in precipitation in Northern Hemisphere mid-latitude regions taken as a whole (see Fig. 4). Stronger trends are generally found over the last four decades. In general, the findings are that wet areas are getting wetter and dry areas are getting drier, consistent with an overall intensification of the hydrological cycle in response to the warming climate (IPCC 2013).

As mentioned earlier, it is well known that warmer air can contain more water vapor than cooler air. Global analyses show that the amount of water vapor in the atmosphere has in fact increased over both land and oceans. Climate change also alters dynamical characteristics of the atmosphere that in turn affect weather patterns and storms. At mid-latitudes, there is an upward trend in extreme precipitation in the vicinity of fronts associated with mid-latitude storms. Locally, natural variations can also be important. In contrast, the subtropics are generally tending to have less overall rainfall and more droughts. Nonetheless, many areas show an increasing tendency for larger rainfall events when it does rain (Janssen et al. 2014; Melillo et al. 2014; IPCC 2013).

Climate Change Effects on Severe Weather

Along with the overall changes in climate, there is strong evidence of an increasing trend over recent decades in some types of extreme weather events, including their frequency, intensity, and duration, with resulting impacts on our society. The changing trends in severe weather resulting from climate change are already

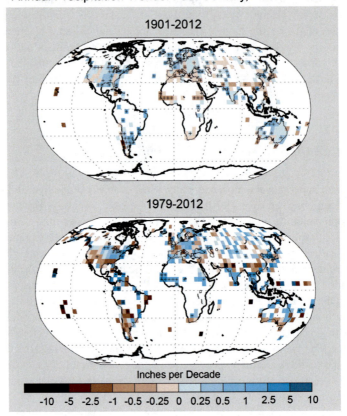

Fig. 4 Global precipitation trends for the period 1901–2012 (top) and 1979–2012 (bottom). Based on date from NOAA NCDC. From Melillo et al. (2014)

affecting the world, including the United States. The United States has sustained over 178 weather/climate disasters since 1980 where damages/costs reached or exceeded $1 billion per event (including CPI adjustment to 2013), with an overall increasing trend (http://www.ncdc.noaa.gov/billions/; also Smith and Katz 2013). The total cost of the 178 events through 2014 is over $1 trillion. In the years 2011 and 2012, there were more such weather events than previously experienced in any given year, with 14 events in 2011 and 11 in 2012, with costs greater than $60 billion in 2011 and greater than $110 Billion during 2012. There were 8 billion dollar plus events in the United States in 2014. The events in these analyses include major heat waves, severe storms, tornadoes, droughts, floods, hurricanes, and wildfires. A portion of these increased costs can be attributed to the increase in population and infrastructure near coastal regions. However, even if hurricanes and their large, mostly coastal, impacts were excluded, there still would be an overall increase in the number of billion dollar events over the last 34 years. Similar analyses by Munich Re and other organizations show that there are growing numbers of severe weather events worldwide causing extensive damage and loss of lives. Figure 5 shows the overall increase in the number of severe events since 1980 through 2015. Even though geophysical events like earthquakes are included in Fig. 5, they are roughly a constant number each year, while the number of severe climate and weather related events has increased dramatically. In summary, there is a clear trend in the impacts of severe weather events on human society not only in the United States, but throughout the world.

Throughout the world, the trends in extreme events are changing; these include increases in the number of extremely hot days, less extreme cold days, more precipitation events coming as unusually large precipitation, and more floods in some regions and more drought in others (Min et al. 2011; IPCC 2012, 2013; Zwiers et al. 2013; Melillo et al. 2014; Wuebbles et al. 2014a, b). For the United States, analyses of atmospheric observations (e.g., Kunkel et al. 2013; Peterson et al. 2013; Vose et al. 2014; Wuebbles et al. 2014a), have shown a pattern of responses in weather extremes relative to the changing climate. These analyses have shown that there are some events, especially those relating to temperature and precipitation extremes, where there is strong understanding of the trends in extreme weather and also of the underlying causes of the observed changes. For some other extremes, the detection of trends in floods, droughts, and extratropical cyclones is also high, but there is less

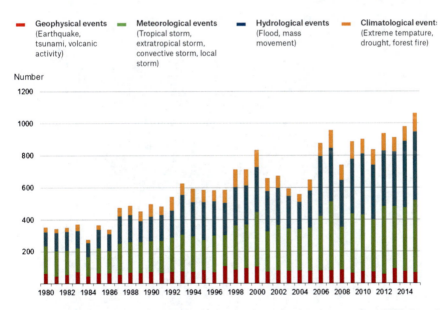

Fig. 5 The number of severe loss events from natural catastrophes per year since 1980 through 2015 as evaluated by Munich Re. Overall losses totaled $90 billion dollars (2015 was not a high year in terms of total costs; the previous year was $110 billion), of which roughly $27 billion was insured. In 2015, natural catastrophes claimed 23,000 lives (average over the last 30 years was 54,000). Figure from Munich Re (https://www.munichre.com/us/weather-resilience-and-protection/media-relations/news/160104-natcatstats2015/index.html)

(only medium) understanding of the underlying cause of the trends. Similarly, there is medium understanding of the observed trends and cause of changes in hurricanes and also in snow events. There is insufficient data to accurately determine trends in strong winds, hail, ice storms, and tornadoes, so there response to a changing climate are not as well understood. Findings for the United States correlate well with analyses of climate extremes globally (IPCC 2012, 2013).

Modeling studies of the changes in climate are generally consistent with the observed trends in extreme weather events over recent decades. Extreme weather events obviously occur naturally. However, the overall changes in climate occurring globally are also altering the frequency and/or severity of many of these extreme events. Trends in extreme weather events, especially in more hot days, less cold days, and more precipitation coming as extreme events, are expected to continue and to intensify over the coming decades.

In most of the United States over the four decades or so, the heaviest rainfall events have become more frequent (e.g., see Fig. 6) and the amount of rain falling in very heavy precipitation events has been significantly above average. This increase has been greatest in the Northeast, Midwest, and upper Great Plains (Melillo et al. 2014).

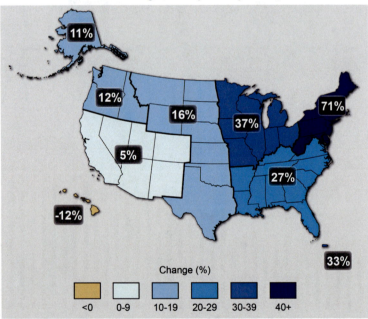

Fig. 6 Percent increases in the amount of precipitation falling in very heavy events (defined as the heaviest 1% of all daily events) from 1958 to 2012 for each region of the continental U.S. These trends are larger than natural variations for the Northeast, Midwest, Puerto Rico, Southeast, Great Plains, and Alaska. The trends are not larger than natural variations for the Southwest, Hawaii, and the Northwest. The changes shown in this figure are calculated from the beginning and end points of the trends for 1958 to 2012. From Melillo et al. (2014)

Similar findings are being found in many other parts of the world. Basic physics tells us that a warmer atmosphere should generally hold more water vapor, so this finding is not so surprising. A number of studies suggest that these trends will continue (Janssen et al. 2014; Melillo et al. 2014; Wuebbles et al. 2014a, b).

Heat waves occur naturally within the climate system—while the timing and location of an individual heat wave may be largely a natural phenomenon, this event can also be affected by human-induced climate change (Trenberth and Fasullo 2012). There is emerging evidence that climate change is affecting most of the increasing heat wave severity over our planet. There has been a detectable human influence for major recent heat waves in the United States (Meehl et al. 2009; Rupp et al. 2012; Duffy and Tebaldi 2012), Europe (Stott et al. 2010; Trenberth 2011), and Russia (Christidis et al. 2011). For example, analyses of the summer 2011 heat wave and drought in Oklahoma and Texas, which cost Texas an estimated $8 billion in agricultural losses, have shown that human-driven climate change approximately doubled the probability that the heat was record-breaking (Hoerling et al. 2013). The possibility of record-breaking temperature extremes has increased and will likely continue to increase as the global climate continues to warm. The changes in climate are thus increasing the likelihood for these types of severe events.

The largest, most damaging, storms are tropical cyclones, referred to as hurricanes when they occur in the Atlantic Ocean. Over the 40 years of satellite monitoring, there has been a shift toward stronger hurricanes in the Atlantic, with fewer smaller (category 1 and 2) hurricanes and more intense (category 4 and 5) hurricanes. A variety of studies have suggested that the intensity of hurricanes should increase under a changing climate but that the overall number of hurricanes may not be affected or possibly even decrease. Observations show no significant trend in the global number of tropical cyclones (IPCC 2012, 2013) nor has any trend been identified in the number of U.S. landfalling hurricanes (Melillo et al. 2014).

Trends remain uncertain in some types of severe weather, including the intensity and frequency of tornadoes, hail, and damaging thunderstorm winds, but such events are under scrutiny to determine if there is a climate change influence. Initial studies do suggest that tornadoes could get more intense in the coming decades (Diffenbaugh et al. 2013).

After at least two thousand years of little change, the world's sea level rose by roughly 0.2 m (8 in.) over the last century, and satellite data provide evidence that the rate of rise over the past 20 years has roughly doubled. Sea level is rising because ocean water expands as it heats up and because water is added to the oceans from melting glaciers and ice sheets. Also, the observed increase in atmospheric carbon dioxide (CO_2) resulting largely from fossil fuel burning also results in increasing the amount of CO_2 in the oceans and thus, a larger amount of carbonic acid. The oceans are currently absorbing about a quarter of the carbon dioxide emitted to the atmosphere annually (Le Quéré et al. 2009) and are becoming more acidic as a result, leading to concerns about intensifying impacts on marine ecosystems (Melillo et al. 2014).

The Basis for a Human Cause for Climate Change

External forcings on the Earth's climate can occur naturally or from the effects of human activities. Natural forcings on climate include variations in energy received from the Sun, the effects of volcanic eruptions, and changes in the Earth's orbit, with associated variations in sunlight across the world. There are also factors that are internal to the climate system that are the result of complex interactions between the atmosphere with the ocean, land surface, and life on Earth. These internal factors include natural modes of climate system variability, such as those that form El Nino events in the Pacific Ocean.

Natural changes in external forcings and internal factors have been entirely responsible for climate changes in the distant past. At the global scale, over multiple decades, the impact of external forcings on temperature far exceeds that of internal variability (which is less than 0.5 °F (Swanson et al. 2009)). At the regional scale, and over shorter time periods, internal variability can be responsible for much larger changes in temperature and other aspects of climate. Today, however, the picture is very different. Although natural factors still affect climate, it is now understood that human activities are the primary cause of the changes in climate for at least the last six decades and perhaps much longer: specifically, human activities that increase atmospheric levels of CO_2 and other heat-trapping gases and various particles that, depending on the type of particle, can have either a heating or cooling influence on climate (Melillo et al. 2014).

The greenhouse effect is key to understanding how human activities affect the Earth's climate. As the Sun shines on the Earth, the Earth heats up. The Earth then re-radiates this heat back to space. Some gases, including H_2O, CO_2, ozone (O_3), methane (CH_4), and nitrous oxide (N_2O), absorb some of the heat given off by the Earth's surface and lower atmosphere. These heat-trapping gases then reradiate the energy, with the result of effectively trapping some of the heat inside the climate system (e.g., see Melillo et al. 2014). This greenhouse effect is a natural process, first proposed in 1824 by the French mathematician and physicist Joseph Fourier and confirmed in laboratory studies by British scientist John Tyndall starting in 1859. The Earth is as we know it because of the greenhouse effect. The Earth would be a frozen planet, about 60 °F colder than today, without this natural greenhouse effect (but assuming the same albedo, or reflectivity, as today).

Over the last five decades, natural drivers of climate such as solar forcing and volcanoes would actually have led to a slight cooling. For example, accurate observations of the Sun from satellites since 1978 show that the solar output has actually decreased slightly from 1978 to now. Natural drivers cannot explain the observed warming over this period. The majority of the warming can only be explained by the effects of human influences (Stott et al., 2010; Gillet et al. 2012; IPCC 2013; Santer et al. 2013), especially the emissions from burning fossil fuels (i.e., coal, oil, and natural gas), and from changes in land use, such as deforestation. As a result of human activities, atmospheric concentrations of various gases and particles are changing, including those for CO_2, CH_4, and N_2O, and particles such as black carbon (soot), which has a warming influence, and sulfates, which have an overall cooling influence (because they reflect sunlight). The most important changes are occurring in the concentration of CO_2; its atmospheric concentration has now reached 400 ppm (400 molecules per 1 million molecules of air; this small amount is important because of the heat-trapping ability of CO_2). 400 ppm of CO_2 has not been seen on Earth for over 1 million years, well before the appearance of humans—preindustrial levels of CO_2 were approximately 280 ppm. The increase in CO_2 over the last several hundred years is almost entirely due to burning of fossil fuels and to a lesser extent, from land use change (IPCC 2013).

The conclusion that human influences are the primary driver of recent climate change is based on multiple lines of independent evidence. The first line of evidence is our fundamental understanding of how certain gases trap heat (these so-called greenhouse gases include H_2O, CO_2, CH_4, N_2O, and some other gases and particles that can all absorb the infrared radiation emitted from the Earth that otherwise would go to space), how the climate system responds to changing concentrations of these gases, and how other factors, both natural and human induced, affect climate.

Also the reconstructions of past climates (e.g., from a variety of datasets including those from tree rings, ice cores, and corals) show that recent changes in global surface temperatures are highly unusual and outside the range of natural variability. These studies show that the last decade (2000–2009) has been much warmer than any period in the last 1300 years and perhaps much longer (IPCC 2013; PAGES 2K Consortium 2013; Mann et al. 2008). Through 2016, it appears that this decade will be much warmer than the previous decade.

The rate of globally averaged surface air temperature increase was slower in the period from 2000 to 2009 than it was in the prior three decades, but such variability is to be expected and does not conflict with the understanding of the processes affecting climate change. This past decade was still the warmest decade in the observational record. Global surface air temperature can be affected by natural variability on the scale of about a decade (for further discussion, see IPCC 2013; Melillo et al. 2014; Karl et al. 2015). Also, other climate change indicators, like the decrease in Arctic sea ice and sea level rise, have not seen a slower change in the rate of change during the same period.

Climate models provide additional evidence through studies to simulate the climate of the past century that separate the human and natural factors that influence climate. As shown in Fig. 7, when the human-related emissions are removed, these models show that natural factors (solar variations and volcanic activity) would have tended to lead to a slight cooling, and other natural variations are too small to explain the observed warming (IPCC 2013). Human influences are the only way to reproduce the temperature increase observed over the past six decades.

21st Century Projections of Climate Change

Climate models have analyzed projections of future conditions under a range of emissions scenarios (that depend on assumptions of population change, economic

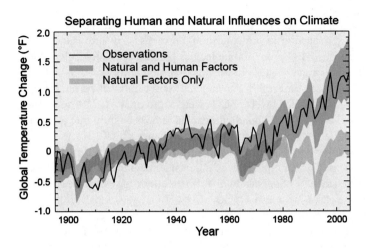

Fig. 7 Observed global average changes (black line), and model simulations using only changes in natural factors (solar and volcanic) in green, and with the addition of human-induced emissions (blue). Climate changes since 1950 cannot be explained by natural factors or variability, and can only be explained by human factors. *Figure source* adapted from Huber and Knutti (2011). From Melillo et al. (2014)

development, our continued use of fossil fuels, changes in other human activities, and other factors). All of the 20+ models used in IPCC (2013) show warming by late this century that is much larger than historical variations nearly everywhere (see Fig. 8). For precipitation, the climate models show decreases in precipitation in the subtropics and increases in precipitation at higher latitudes. As discussed earlier, extreme weather events associated with extremes in temperature and precipitation are likely to continue and to intensify.

Choices made now and in the next few decades about emissions from fossil fuel use and land use change will determine the amount of additional future warming over this century and beyond. Global emissions of CO_2 and other heat-trapping gases continue to rise. Climate changes over the rest of this century and beyond depend primarily on the extent of human activities and resulting emissions; and the sensitivity of the climate system to those changes (that is, the response of global temperature to a change in radiative forcing caused by human emissions).

Important factors in future emissions include growth in the economy, the types of energy used, and the future efficiency of cities, buildings, and vehicles; these limit the ability to accurately project future changes in climate. Thus a range of plausible projections of what might happen, under a given set of assumptions, are used. These scenarios describe possible futures in terms of population, energy sources, technology, heat-trapping gas emissions, atmospheric levels of carbon dioxide, and/or global temperature change.

A certain amount of climate change is inevitable as the CO_2 concentration increases in the atmosphere. There is a lag in the response in the Earth's climate system due to the large heat capacity of the oceans and other factors. An additional 0.2–0.3 °C

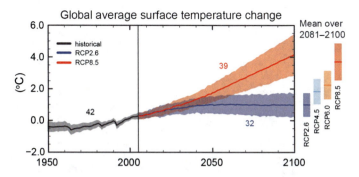

Fig. 8 Multi-model simulated time series from 1950 to 2100 for the change in global annual mean surface temperature relative to 1986–2005 for a range of future emissions scenarios that account for the uncertainty in future emissions from human activities [as analyzed with the 20 + models from around the world used in the most recent international assessment (IPCC 2013)]. The mean and associated uncertainties [1.64 standard deviations (5–95%) across the distribution of individual models (shading)] based on the averaged over 2081–2100 are given for all of the RCP scenarios as colored vertical bars. The numbers of models used to calculate the multi-model mean is indicated. (Figure 7a from IPCC (2013) Summary for Policymakers)

(about 0.5 °F) increase in temperature is inevitable over the next few decades (Matthews and Zickfeld 2012), although natural variability could also be important role on these time scales (Hawkins and Sutton 2011). Higher emissions of CO_2 and other heat-trapping gases would be expected to result in larger climate changes expected by mid-century and beyond. By the second half of the century, uncertainty in what will be the level of future emissions from human activities becomes increasingly dominant in determining the magnitude and patterns of future change, particularly for temperature-related aspects (Hawkins and Sutton 2009, 2011).

A range of future scenarios are examined in Figs. 8 and 9 that vary from assuming strong continued dependence on fossil fuels in energy and transportation systems over the 21st century (scenario RCP8.5) to assuming major mitigation actions (RCP2.6). In all cases, global surface temperature change for the end of the 21st century is *likely* to exceed an increase of 1.5 °C (2.7 °F) relative to the period from 1850 to 1900 for all projections, with the exception of the RCP2.6 scenario (IPCC 2013). The RCP2.6 scenario has much lower effects on climate than the other scenarios because it assumes both significant mitigation to reduce emissions and also that technologies are developed that can remove CO_2 from the atmosphere (thus achieving net negative carbon dioxide emissions).

A number of research studies have examined the potential criteria for dangerous human interferences in climate where it will be difficult to adapt to the changes in climate without major effects on our society (e.g., Hansen et al. 2007). These studies have generally concluded that an increase in globally average temperature of roughly 1.5 °C (2.7 °F) is an approximate threshold for dangerous human interferences with the climate system (see IPCC 2013, 2014 for further discussion; earlier studies had proposed 2 °C). However, this threshold is not exact and the changes in climate vary geographically and resulting impacts are sector dependent.

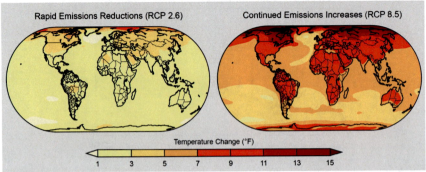

Fig. 9 Projected change in average annual temperature over the period 2071–2099 (compared to the period 1971–2000) under a low scenario that assumes rapid reductions in emissions and concentrations of heat-trapping gases (RCP2.6), and a higher scenario that assumes continued increases in emissions (RCP8.5). From Melillo et al. (2014)

The warming and other changes in the climate system will continue beyond 2100 under all RCP scenarios, except for a leveling of temperature under RCP2.6. In addition, it is fully expected that the warming will continue to exhibit interannual-to-decadal variability and will not be regionally uniform.

Projections of future changes in precipitation show small increases in the global average but substantial shifts in where and how precipitation falls (see Fig. 10). Generally, areas closest to the poles are projected to receive more precipitation, while the dry subtropics (the region just outside the tropics, between 23° and 35° on either side of the equator) will generally expand toward the poles and receives less rain. Increases in tropical precipitation are projected during rainy seasons (such as monsoons), especially over the tropical Pacific. Certain regions, including the western U.S. [especially the Southwest (Melillo et al. 2014) and the Mediterranean (IPCC 2013)], are presently dry and are expected to become drier. The widespread trend of increasing heavy downpours is expected to continue, with precipitation becoming more intense (Gutowski et al. 2007; Boberg et al. 2009; Sillmann et al. 2013). The patterns of the projected changes of precipitation do not contain the spatial details that characterize observed precipitation, especially in mountainous terrain, because of model uncertainties and their current spatial resolution (IPCC 2013).

As mentioned earlier, some areas both in the United States and throughout the world are already experiencing climate-related changes in trends for extreme weather events. These trends are likely to continue throughout this century and perhaps beyond (depending on the actions we take). The following trends are expected based on the existing science understanding over the coming decades (see Melillo et al. 2014, or IPCC 2013, for more details):

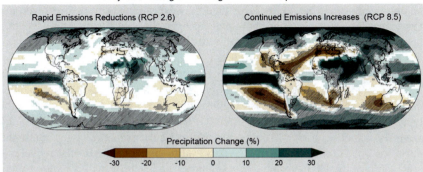

Fig. 10 Projected change in average annual precipitation over the period 2071–2099 (compared to the period 1971–2000) under a low scenario that assumes rapid reductions in emissions and concentrations of heat-trapping gasses (RCP2.6), and a higher scenario that assumes continued increases in emissions (RCP8.5). Hatched areas indicate confidence that the projected changes are significant and consistent among models. White areas indicate that the changes are not projected to be larger than could be expected from natural variability. In general, northern parts of the U.S. (especially the Northeast and Alaska) are projected to receive more precipitation, while southern parts (especially the Southwest) are projected to receive less. From Melillo et al. (2014)

- It is likely that over the coming decades the frequency of warm days and warm nights will increase in most land regions, while the frequency of cold days and cold nights will decrease. As a result, an increasing tendency for heat waves is likely in many regions of the world.
- Some regions are likely to see an increasing tendency for droughts while others are likely to see an increasing tendency for floods. This roughly corresponds to the wet getting wetter and the dry getting drier.
- It is likely that the frequency and intensity of heavy precipitation events will increase over land. These changes are primarily driven by increases in atmospheric water vapor content, but also affected by changes in atmospheric circulation.
- Tropical storm (hurricane)-associated storm intensity and rainfall rates are projected to increase as the climate continues to warm.
- Initial studies also suggest that tornadoes are likely to become more intense. However, this is more uncertain.
- For some types of extreme events, like wind storms, and ice and hail storms, there is too little understanding currently of how they will be affected by the changes in climate.

Around the world, many millions of people and many assets related to energy, transportation, commerce, and ecosystems are located in areas at risk of coastal flooding because of sea level rise and storm surge. Sea level is projected to rise an additional 0.3–1.2 m (1–4 ft) in this century (see Fig. 11; Melillo et al. 2014; similar findings in IPCC 2013). The best estimates for the range of sea level rise projections for this century remain quite large; this may be due in part to what emissions scenario we follow, but more importantly it depends on just how much melting occurs from the ice on large land masses, especially from Greenland and Antarctica. Recent projections show that for even the lowest emissions scenarios, thermal expansion of ocean waters (Yin 2012) and the melting of small mountain glaciers (Marzeion et al. 2012) will result in 11 in. of sea level rise by 2100, even without any contribution from the ice sheets in Greenland and Antarctica. This suggests that about 0.3 m (1 ft) of global sea level rise by 2100 is probably a realistic low end. Recent analyses suggest that 1.2 m (4 ft) may be a reasonable upper limit (Rahmstorf et al. 2012; IPCC 2013; Melillo et al. 2014). Although scientists cannot yet assign likelihood to any particular scenario, in general, higher emissions scenarios would be expected to lead to higher amounts of sea level rise.

Because of the warmer global temperatures, sea level rise will continue beyond this century. Sea levels will likely continue to rise for many centuries at rates equal to or higher than that of the current century. Many millions of people live within areas than can be affected by the effects of storm surge within a rising sea level. The Low Elevation Coastal Zone (less than 10 m elevation) constitutes 2% of the world's land area, yet contains 10% of the world's population (over 600 million people) (McGranahan et al. 2007; Neumann et al. 2015). Most of the world's megacities are within the coastal zone. By 2030, with sea level rise, the area will expand and 800–900 million people will be exposed (Güneralp et al. 2015; Neumann et al. 2015).

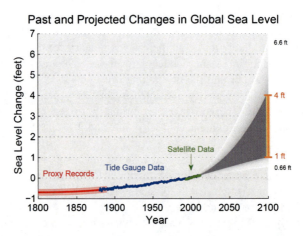

Fig. 11 Estimated, observed, and projected amounts of global sea level rise from 1800 to 2100, relative to the year 2000. Estimates from proxy data (for example, based on sediment records) are shown in red (1800–1890, pink band shows uncertainty), tide gauge data in blue for 1880–2009 (Church and White 2011; Church et al. 2011) and satellite observations are shown in green from 1993 to 2012 (Nerem et al. 2010). The future scenarios range from 0.66 to 6.6 ft in 2100 (Parris et al. 2012). These scenarios are not based on climate model simulations, but rather reflect the range of possible scenarios based on scientific studies. The orange line at right shows the currently projected range of sea level rise of 1–4 ft by 2100, which falls within the larger risk-based scenario range. The large projected range reflects uncertainty about how glaciers and ice sheets will react to the warming ocean, the warming atmosphere, and changing winds and currents. As seen in the observations, there are year-to-year variations in the trend. From Melillo et al. (2014)

As mentioned earlier, CO_2 is dissolving into the oceans where it reacts with seawater to form carbonic acid, lowering ocean pH levels ("acidification") and threatening a number of marine ecosystems (Doney et al. 2009). The oceans have absorbed 560 billion tons of CO_2 over the last 250 years, thus increasing the acidity of surface waters by 30% (Melillo et al. 2014). The current observed rate of change is roughly 50 times faster than known historical change (Hönisch et al. 2012; Orr 2011; Caldeira and Wickett 2003). Ocean acidification hotspots are occurring due to regional factors such as coastal upwelling (Feely et al. 2008), changes in discharge rates from rivers and glaciers (Mathis et al. 2011) sea ice loss (Yamamoto-Kawai et al. 2009), and urbanization (Feely et al. 2010).

The acidification of the oceans is suppressing carbonate ion concentrations that are critical for marine calcifying animals such as corals, zooplankton, and shellfish. Many of these animals form the foundation of the marine food web. Today, more than a billion people worldwide rely on food from the ocean as their primary source of protein. Ocean acidification puts this important resource at risk.

Higher emission scenario projections could reduce the ocean pH from the current 8.1 to as low as 7.8 by the end of the century (Orr et al. 2005). This is unprecedented in human history—such large rapid changes in ocean pH have probably not been experienced for the past 100 million years, and it is unclear whether and how quickly ocean life could adapt to such rapid acidification (Hönisch et al. 2012). Potential impacts on food supplies from the oceans are unclear. Unfortunately, since sustained efforts to monitor ocean acidification worldwide are only beginning, it is currently impossible to quantify this risk or to be able to predict exactly how ocean acidification impacts will cascade throughout the marine food chain and affect the overall structure of marine ecosystems.

Responding to Climate Change: A Look Forward

It has become increasingly clear that our future depends on how we act to limit climate change. Science is the basis for developing responses to climate change, by providing the:

- Motivation for seeking to develop a cost-effective plan to reduce those impacts;
- Sense of urgency for doing so now rather than waiting;
- Awareness that such a plan must include both mitigation and adaptation;
- Knowledge of the sources of the offending emissions and the character of society's vulnerabilities that allows appropriate specificity in designing a plan; and
- Recognition that any U.S. plan must include a component designed to bring other countries along.

We basically have three choices:

- Mitigation, meaning measures to reduce the pace and magnitude of the changes in global climate being caused by human activities.

- Adaptation, meaning measures to reduce the adverse impacts on human well-being resulting from the changes in climate that do occur.
- Suffering the adverse impacts and societal disruption that are not avoided by either mitigation or adaptation.

Right now we are doing some of all three. What's up for grabs is the future mix. Minimizing the amount of suffering in that mix can only be achieved by doing a lot of mitigation and a lot of adaptation. Mitigation alone would be inadequate; climate is already changing and can't be stopped quickly. Adaptation alone would also be inadequate; adaptation gets costlier and less effective as climate change grows.

We must reduce emissions of the heat-trapping gases and particles to avoid unmanageable levels of climate change and the resulting impacts. At the same time we need to adapt to the changes in climate that are unavoidable. Adaptation is not a choice—our choice is whether to adapt proactively or respond to the consequences. Adaptation requires a paradigm shift, focusing on managing risks. Proactively preparing for climate change can reduce impacts while also facilitating a more rapid and efficient response to changes as they happen. Such efforts are beginning in the United States and other parts of the world, to build adaptive capacity and resilience to climate change impacts. Using scientific information to prepare for climate changes in advance can provide economic opportunities, and proactively managing the risks can reduce impacts and costs over time.

In the United States, the first major steps were taken on June 25, 2013, when President Obama announced the Climate Action Plan, a national plan for tackling climate change. The plan, is divided into three sections that outline steps to (1) cut carbon pollution in the United States, including standards for both new and existing power plants, (2) actions to prepare the United States for the impacts of climate change, and (3) plans to lead international efforts to address global climate change. Also, the President's Climate Action Plan fast-tracks permitting for renewable energy projects on public lands, increases funding for clean energy technology and efficiency improvements, and calls for improved efficiency standards for buildings and appliances, as well as heavy trucks. The plan additionally establishes the first-ever Federal Quadrennial Energy Review to encourage strategic national energy planning. As part of the plan, the American Business Act on Climate Pledge has received commitments from 154 companies (so far) from across the American economy for their contributions to mitigation and adaptation. Agreements made with China, India, and other countries have been important in getting to an international agreement on climate change.

Large reductions in global emissions of heat-trapping gases will be important if we are to reduce the risks associated with many of the worst impacts of climate change. The international agreement made in Paris by 195 countries in December 2015 is an important start to achieving this. The 21st annual Conference of Parties (COP21) resulted in a global action plan to reduce emissions of carbon dioxide and other greenhouse gases. The current Paris Agreement only extends through 2030 but the long term goal is to keep the increase in global average temperature to well below 2 °C (3.6 °F) above pre-industrial levels. This itself will be extremely difficult to do, but the ultimate aim would be to keep the temperature change below 1.5 °C (2.7 °F). This would be roughly equivalent to following the extremely low RCP2.6 scenario discussed earlier (about half of the global climate models used in the 2013 IPCC assessment produced a change of about 1.5 °C).

The current agreement is not sufficient to reach even the 2 °C limit but it is an important step towards getting there and perhaps to 1.5 °C. Its full implementation throughout the world, including the United States, should lead to incentives for the development of new energy and transportation technologies that should further reduce emissions. This is an important step. It is clear that the choices we make to reduce climate change over the next few decades will not only affect us, they will affect our children, our grandchildren, and future generations.

References

Boberg F, Berg P, Thejll P, Gutowski WJ, Christensen JH (2009) Improved confidence in climate changeprojections of precipitation evaluated using daily statistics from the PRUDENCE ensemble. Clim Dyn 32:1097–1106. https://doi.org/10.1007/s00382-008-0446-y

Caldeira K, Wickett ME (2003) Oceanography: anthropogenic carbon and ocean pH. Nature 425:365. https://doi.org/10.1038/425365a

Christidis N, Stott PA, Brown SJ (2011) The role of human activity in the recent warming of extremely warm daytime temperatures. J Clim 24:1922–1930. https://doi.org/10.1175/2011JCLI4150.1

Church JA, White NJ (2011) Sea-level rise from the late 19th to the early 21st century. Surv Geophys 32:585–602. https://doi.org/10.1007/s10712-011-9119-1

Church JA, White NJ, Konikow LF, Domingues CM, Cogley JG, Rignot E, Gregory JM, van den Broeke MR, Monaghan AJ, Velicogna I (2011) Revisiting the Earth's sea-level and energy budgets from 1961 to 2008. Geophys Res Lett 38:L18601. https://doi.org/10.1029/2011GL048794

Deser C, Knutti R, Solomon S, Phillips AS (2012) Communication of the role of natural variability in future North American climate. Nat Clim Change 2:775–779. https://doi.org/10.1038/nclimate1562

Diffenbaugh NS, Scherer M, Trapp RJ (2013) Robust increases in severe thunderstorm environments in response to greenhouse forcing. Proc Natl Acad Sci 110:16361–16366. https://doi.org/10.1073/pnas.1307758110

Doney SC, Fabry VJ, Feely RA, Kleypas JA (2009) Ocean acidification: the other CO_2 problem. Ann Rev Mar Sci 1:169–192. https://doi.org/10.1146/annurev.marine.010908.163834

Duffy PB, Tebaldi C (2012) Increasing prevalence of extreme summer temperatures in the U.S. Clim Change 111:487–495. https://doi.org/10.1007/s10584-012-0396-6

Feely RA, Sabine CL, Hernandez-Ayon JM, Ianson D, Hales B (2008) Evidence for upwelling of corrosive "acidified" water onto the continental shelf. Science 320:1490–1492. https://doi.org/10.1126/science.1155676

Feely RA, Alin SR, Newton J, Sabine CL, Warner M, Devol A, Krembs C, Maloy C (2010) The combined effects of ocean acidification, mixing, and respiration on pH and carbonate saturation in an urbanized estuary. Estuar Coast Shelf Sci 88:442–449. https://doi.org/10.1016/j.ecss.2010.05.004

Gillett NP, Arora VK, Flato GM, Scinocca JF, Salzen KV (2012) Improved constraints on 21st-century warming derived using 160 years of temperature observations. Geophys Res Lett 39:L01704. https://doi.org/10.1029/2011GL050226

Güneralp B, Güneralp İ, Liu Y (2015) Changing global patterns of urban exposure to flood and drought hazards. Glob Environ Change 31:217–225

Gutowski WJ, Takle ES, Kozak KA, Patton JC, Arritt RW, Christensen JH (2007) A possible constraint on regional precipitation intensity changes under global warming. J Hydrometeorology 8:1382–1396. https://doi.org/10.1175/2007jhm817.1

Hansen J, Sato M et al (2007) Dangerous human-made interference with climate: a GISS modelE study. Atmos Chem Phys 7:2287–2312. https://doi.org/10.5194/acp-7-2287-2007

Hawkins E, Sutton R (2009) The potential to narrow uncertainty in regional climate predictions. Bull Am Meteor Soc 90:1095–1107. https://doi.org/10.1175/2009BAMS2607.1

Hawkins E, Sutton R (2011) The potential to narrow uncertainty in projections of regional precipitation change. Clim Dyn 37:407–418

Hoerling M, Chen M, Dole R, Eischeid J, Kumar A, Nielsen-Gammon JW, Pegion P, Perlwitz J, Quan X-W, Zhang T (2013) Anatomy of an extreme event. J Clim 26:2811–2832. https://doi.org/10.1175/JCLI-D-12-00270.1

Hönisch B, Ridgwell A et al (2012) The geological record of ocean acidification. Science 335:1058–1063. https://doi.org/10.1126/science.1208277

Huber M, Knutti R (2011) Anthropogenic and natural warming inferred from changes in Earth's energy balance. Nature Geosci. 5:31–36. https://doi.org/10.1038/ngeo1327

IPCC (Intergovernmental Panel on Climate Change) (2012) Managing the risks of extreme events and disasters to advance climate change adaptation. In: Field CB, Barros V, Stocker TF, Qin D, Dokken DJ, Ebi KL, Mastrandrea MD, Mach KJ, Plattenr G-K, Allen SK, Tignor M, Midgley PM (eds) A special report of the intergovernmental panel on climate change. Cambridge University Press, Cambridge, United Kingdom, p 582

IPCC (Intergovernmental Panel on Climate Change) (2013) Climate change 2013: the physical science basis. In: Stocker TF, Qin D, Plattner G-K, Tignor M, Allen SK, Boschung J, Nauels A, Xia Y, Bex V, Midgley PM (eds) Contribution of working group I to the fifth assessment report of the intergovernmental panel on climate change. Cambridge University Press, Cambridge, United Kingdom and New York, USA

IPCC (Intergovernmental Panel on Climate Change) (2014) Climate change 2014: synthesis report. In: Core Writing Team, Pachauri RK, Meyer LA (eds) Contribution of working groups I, II and III to the fifth assessment report of the intergovernmental panel on climate change. IPCC, Geneva, Switzerland, p 151

Janssen E, Wuebbles DJ, Kunkel KE, Olsen SC, Goodman A (2014) Trends and projections of extreme precipitation over the contiguous United States. Earth's Future 2:99–113. https://doi.org/10.1002/2013EF000185

Karl TR, Arguez A, Huang B, Lawrimore JH, McMahon JR, Menne MJ, Peterson TC, Vose RS, Zhang H (2015) Possible artifacts of data biases in the recent global surface warming hiatus. Science 348:1469–1472

Kunkel KE et al (2013) Monitoring and understanding changes in extreme storm statistics: state of knowledge. Bullet Am Meteorol Soc 94:499–514. https://doi.org/10.1175/BAMS-D-11-00262.1

Le Quéré C et al (2009) Trends in the sources and sinks of carbon dioxide. Nat Geosci 2:831–836. https://doi.org/10.1038/ngeo689

Mann ME, Zhang Z, Hughes MK, Bradley RS, Miller SK, Rutherford S, Ni F (2008) Proxy-based reconstructions of hemispheric and global surface temperature variations over the past two millennia. Proc Natl Acad Sci 105:13252–13257. https://doi.org/10.1073/pnas.0805721105

Marzeion B, Jarosch AH, Hofer M (2012) Past and future sea level change from the surface mass balance of glaciers. Cryosphere Discuss 6:3177–3241. https://doi.org/10.5194/tcd-6-3177-2012

Mathis JT, Cross JN, Bates NR (2011) Coupling primary production and terrestrial runoff to ocean acidification and carbonate mineral suppression in the eastern Bering Sea. J Geophys Res 116: C02030. https://doi.org/10.1029/2010JC006453

Matthews HD, Zickfeld K (2012) Climate response to zeroed emissions of greenhouse gases and aerosols. Nat Clim Change 2:338–341. https://doi.org/10.1038/nclimate1424

McGranahan G, Balk D, Anderson B (2007) The rising tide: assessing the risks of climate change and human settlements in low elevation coastal zones. Environ Urbanization 19:17–37

Meehl GA, Tebaldi C, Walton G, Easterling D, McDaniel L (2009) Relative increase of record high maximum temperatures compared to record low minimum temperatures in the U.S. Geophys Res Lett 36:L23701. https://doi.org/10.1029/2009GL040736

Melillo JM, Richmond TC, Yohe GW (eds) (2014) Climate change impacts in the United States: the third national climate assessment. In: U.S. global change research program, p 840. Available at http://nca2014.globalchange.gov

Min S, Zhang X, Zwiers F, Hegerl G (2011) Human contribution to more-intense precipitation extremes. Nature 470:378–381

Nerem RS, Chambers DP, Choe C, Mitchum GT (2010) Estimating mean sea level change from the TOPEX and Jason altimeter missions. Mar Geodesy 33:435–446. https://doi.org/10.1080/01490419.2010.491031

Neumann B, Vafeidis AT, Zimmermann J, Nicholls RJ (2015) Future coastal population growth and exposure to sea-level rise and coastal flooding—a global assessment. PLoS ONE 10(3): e0118571. https://doi.org/10.1371/journal.pone.0118571

Orr JC (2011) Recent and future changes in ocean carbonate chemistry. In: Gattuso J-P, Hansson L (eds) Ocean acidification. Oxford University Press, Oxford, pp 41–66

Orr JC et al (2005) Anthropogenic ocean acidification over the twenty-first century and its impact on calcifying organisms. Nature 437:681–686. https://doi.org/10.1038/nature04095

PAGES 2 K Consortium (2013) Continental-scale temperature variability during the past two millennia. Nat Geosci 6:339–346. https://doi.org/10.1038/ngeo1797

Parris A, Bromirski P, Burkett V, Cayan D, Culver M, Hall J, Horton R, Knuuti K, Moss R, Obeysekera J, Sallenger A, Weiss J (2012) Global sea level rise scenarios for the United States national climate assessment. In: NOAA tech memo OAR CPO-1, National oceanic and atmospheric administration, pp 37. Available online at http://scenarios.globalchange.gov/sites/default/files/NOAA_SLR_r3_0.pdf

Peterson TC et al (2013) Monitoring and understanding changes in heat waves, cold waves, floods and droughts in the United States: state of knowledge. Bull Am Meteorol Soc 94:821–834. https://doi.org/10.1175/BAMS-D-12-00066.1

Rahmstorf S, Perrette M, Vermeer M (2012) Testing the robustness of semi-empirical sea level projections. Clim Dyn 39:861–875. https://doi.org/10.1007/s00382-011-1226-7

Rupp DE, Mote PW, Massey N, Rye CJ, Jones R, Allen MR (2012) Did human influence on climate make the 2011 Texas drought more probable? In: Peterson TC, Stott PA, Herring S (eds) Explaining extreme events of 2011 from a climate perspective. Bulletin American Meteorological Society, pp 1052–1054

Santer BD et al (2013) Identifying human influences on atmospheric temperature. Proc Natl Acad Sci 110:26–33. https://doi.org/10.1073/pnas.1210514109

Sillmann J, Kharin VV, Zwiers FW, Zhang X, Bronaugh D (2013) Climate extremes indices in the CMIP5 multimodel ensemble: part 2. Future climate projections. J Geophys Res Atmos 118:2473–2493. https://doi.org/10.1002/jgrd.50188

Smith AB, Katz RW (2013) U.S. Billion-dollar weather and climate disasters: data sources, trends, accuracy and biases. Nat Hazard 67:387–410

Stott PA, Gillett NP, Hegerl GC, Karoly DJ, Stone DA, Zhang X, Zwiers F (2010) Detection and attribution of climate change: a regional perspective. Wiley Interdisc Rev Clim Change 1:192–211. https://doi.org/10.1002/wcc.34

Swanson KL, Sugihara G, Tsonis AA (2009) Long-term natural variability and 20th century climate change. Proc Natl Acad Sci USA 106:16120–16123. https://doi.org/10.1073/pnas.0908699106

Trenberth KE (2011) Attribution of climate variations and trends to human influences and naturalvariability. Wiley Interdisciplinary Reviews: Climate Change 2:925-930. https://doi.org/10.1002/wcc.142

Trenberth KE, Fasullo JT (2012) Climate extremes and climate change: the Russian heat wave and other climate extremes of 2010. J. Geophys Res Atmos 117:D17103. https://doi.org/10.1029/2012JD018020

UK Royal Society (UKRS) and U.S. National Academy of Sciences (NAS) (2014) Climate change: evidence and causes. National Academy Press, Washington, D.C.

Vose RS, Applequist S, Menne MJ, Williams CN Jr, Thorne P (2012) An intercomparison of temperature trends in the US historical climatology network and recent atmospheric reanalyses. Geophys Res Lett 39, L10703, https://doi.org/10.1029/2012gl051387

Vose RS et al (2014) Monitoring and understanding changes in extremes: extratropical storms, winds, and waves. Bull Am Meteor Soc. https://doi.org/10.1175/BAMS-D-12-00162.1

Wuebbles DJ et al (2014a) CMIP5 climate model analyses: climate extremes in the United States. Am Meteorol Soc, Bullet. https://doi.org/10.1175/BAMS-D-12-00172.1

Wuebbles DJ, Kunkel K, Wehner M, Zobel Z (2014b) Severe weather in the United States under a changing climate. EOS 95:149–150. https://doi.org/10.1002/2014EO180001

Yamamoto-Kawai M, McLaughlin FA, Carmack EC, Nishino S, Shimada K (2009) Aragonite undersaturation in the Arctic Ocean: effects of ocean acidification and sea ice melt. Science 326:1098–1100. https://doi.org/10.1126/science.1174190

Yin J (2012) Century to multi-century sea level rise projections from CMIP5 models. Geophys Res Lett 39:L17709. https://doi.org/10.1029/2012GL052947

Zwiers F, Alexander L, Hegerl G, Knutson T, Kossin J, Naveau P, Nicholls N, Schaar C, Seneviratne S, Zhang X (2013) Climate extremes: challenges in estimating and understanding recent changes in the frequency and intensity of extreme climate and weather events. In: Asrar G, Hurrell J (eds) Climate science for serving society. Springer, Netherlands, pp 339–389

Chapter 3
Cumulative Harm as a Function of Carbon Emissions

John Nolt

Abstract Anthropogenic climate change is indisputably harmful. Yet the nature, extent, and duration of this harm, and hence the extent of our individual and collective causal responsibility for it, are underappreciated. Climate disruption may persist for millennia, during which its harmful effects, lessened by adaptation and moderating temperatures, might diminish in frequency but will nevertheless continue to accumulate in number. Total harm can therefore be meaningfully estimated only relative to some specified time period. With regard to varying emissions scenarios, harm during any such time period within the next few millennia increases directly and (to a close approximation) continuously with cumulative prior anthropogenic carbon emissions up through that period. It follows that even small emissions can in the long run cause significant harm.

Introduction

Anthropogenic climate change is indisputably harmful. Yet the nature, extent, and duration of this harm, and hence the depth of our individual and collective responsibility for it, are underappreciated. Climate disruption may persist for millennia, during which its harmful effects, lessened by adaptation and moderating temperatures, might diminish in frequency but will nevertheless continue to accumulate in number. We can assign no definite date to its cessation. Its harms can therefore be meaningfully estimated only relative to some specified time period. This essay contends that with regard to varying emissions scenarios, harm during any time period p within the next few millennia increases directly and (to a close approximation) continuously with cumulative anthropogenic carbon emissions up through p. It follows that even small emissions can in the long run cause significant harm.

J. Nolt (✉)
Philosophy Department, University of Tennessee, Knoxville, USA
e-mail: nolt@utk.edu

It is sometimes said by way of rationalization that our emissions are justified by the benefits (to us and to the future) of the activities that produce them. But that is clearly true only for emissions that are morally necessary (Nolt 2011b, 2013b; Shue 1993). Morally necessary emissions are those that are not eliminable without violating overriding moral obligations—for example, emissions necessary to prevent dire poverty or loss of life. There are, of course, plenty of borderline cases, and these are subject to reasonable dispute. But many of the uses now made of greenhouse gas emitting technologies in rich nations, or by the rich in poor nations, are plainly not morally necessary. Consider, for example, the burning of fossil fuels for entertainment, for the heating and cooling of trophy houses, for wasteful lighting, or for transportation in overlarge and overpowered vehicles. Given that (as is shown below) such emissions contribute to the bodily harms of others, their moral permissibility is dubious. (For more on morally necessary and morally unnecessary uses of greenhouse-gas-emitting devices, see Nolt 2013b: 149–151.)

But this essay is not about moral necessity or moral justification. Its aim is merely to lay out in greater detail than I have elsewhere (Nolt 2013b, 2015a: 14–15) how harm depends on cumulative emissions and to consider what this means for *causal* (not ethical) responsibility. An action is *causally* responsible for an event if it is among the causes of that event, contributing to its occurrence or intensity. One can be causally but not ethically responsible for an event—as, for instance, when an innocently unknowing carrier of a communicable disease infects someone else. But a detailed account of the conditions of moral responsibility is beyond the scope of this paper.

My argument, in brief, is as follows. Choose any time period p over the next few millennia and consider how various emissions scenarios would affect global average temperature and levels of harm during that period. Then it is true to a close approximation that:

Global average temperature during p increases directly and continuously with our cumulative carbon emissions from now through the end of p, and
Harm to human and nonhuman life during p increases directly and continuously with global average temperature during p.

Therefore,

Harm to human and nonhuman life during p increases directly and continuously with our cumulative carbon emissions up through p.

In what follows I'll explain each premise, comment on the argument as a whole, consider some objections, and, finally, examine an important implication: that even small emissions can cause significant harm to humans in the form of injury, illness, displacement and death—and to non-human life.

I focus on carbon emissions, rather than greenhouse gas emissions generally, for two reasons. First, CO_2 released by various human activities is responsible for about 76% of anthropogenic climate change; and, second, it has a much longer atmospheric lifetime than the second worst culprit, methane, which is responsible for about 16% (IPCC 2014 Synthesis Report Fig. 1.6, p. 46). Long-term climate disruption, which is the primary concern here, is therefore mainly a result of CO_2 emissions.

CO$_2$ emissions are measured by weight in two distinct ways: metric tons of CO$_2$ and metric tons of elemental carbon, C. To convert metric tons C into metric tons CO$_2$, multiply by 3.67. With this conversion factor in mind, I use the terms "carbon emissions" and "CO$_2$ emissions" interchangeably when exact amounts are not at issue. (Another widely used measure is metric tons CDE (Carbon Dioxide Equivalent), which for CO$_2$ is just metric tons of CO$_2$. But this measure is used for other greenhouse gases too, via formulas that convert quantities of those gases into quantities of CO$_2$ with similar warming potential.)

Temperature as a Function of Cumulative Carbon Emissions

Consider the global average temperature over some fixed time period p during the next few millennia under varying carbon emissions scenarios. This section aims to show that if *per impossibile* other variables were held constant, then that temperature would increase directly and continuously with increases in our cumulative carbon emissions from now through p. (This is the first premise of the argument outlined above.) To see clearly what it means, we must examine the causal relationship between cumulative emissions and temperature.

Atmospheric warming occurs when greenhouse gases absorb infrared radiation (radiant heat) from the earth's surface that would otherwise have escaped into space. Each CO$_2$ molecule can individually absorb packets of this radiation in the form of photons, which increase its kinetic energy. The kinetic energy of large aggregates of molecules is what we measure as heat. Ultimately, this atmospheric heat is dispersed into the environment, most of it being transferred to and stored in the oceans.

There is no significant time lag between emission and atmospheric warming. Each molecule begins to trap and disperse minute amounts of heat as soon as it enters the atmosphere. Because this process of absorption and dispersal is mechanical, not chemical, the composition of the molecule is not altered by it. CO$_2$ molecules continue to trap heat for as long as they remain in the atmosphere.

That can be a long time, but it is not forever. Most of the CO$_2$ molecules are eventually either taken up by photosynthetic organisms or dissolved into the oceans, where they contribute to acidification. Some of the carbon ultimately winds up in soil or silt. Yet once atmospheric CO$_2$ levels are raised, they remain elevated for centuries. In a comprehensive review of the literature on the atmospheric lifetime of fossil fuel carbon dioxide, Archer et al. (2009) reported that "The models agree that 20–35% of the CO$_2$ remains in the atmosphere after equilibration with the ocean (2–20 centuries)." This does not mean that individual CO$_2$ molecules stay in the atmosphere that long. Their paths through the carbon cycle are complex. But it does mean that CO$_2$ *levels* remain significantly elevated for centuries.

This long persistence implies that the ultimate maximum global average temperature reached as a result of our emissions will be largely insensitive to their rate and timing. Maximum warming relative to pre-industrial times will instead be roughly proportional to cumulative total anthropogenic carbon emissions. What matters most in the long run, in other words, is how much total carbon we emit into the atmosphere before the fossil fuel era ends (Stocker 2013; Allen et al. 2009).

Once CO_2 emissions cease, there will be little further temperature increase (Matthews and Solomon 2013). But slow climate feedback processes, nearly all of them positive, insure that—barring geo-engineering or other major disruptions of natural processes—high temperatures will far outlast elevated CO_2 levels. Concluding a survey of the relevant literature, Archer et al. (2009, p. 131) write:

> Nowhere in these model results or in the published literature is there any reason to conclude that the effects of CO_2 release will be substantially confined to just a few centuries. In contrast, generally accepted modern understanding of the global carbon cycle indicates that climate effects of CO_2 releases to the atmosphere will persist for tens, if not hundreds, of thousands of years into the future.

Richard Zeebe (2013) is more specific. He estimates that surface temperatures for a high-end total fossil fuel input of 2.5 trillion metric tons of carbon over 500 years, assuming a mid-range equilibrium climate sensitivity of 3 °C, will remain elevated for 23,000–165,000 years.

Of course, global average atmospheric temperature is influenced by many factors other than our carbon emissions. These include, but are not limited to, volcanic eruptions, which shade and thus cool the atmosphere; solar irradiance, which fluctuates, alternately supplying the earth with slightly more or slightly less heat; and variations in heat exchange between the atmosphere and oceans, such as La Niña events (during which an upwelling of cold water in the Pacific cools the atmosphere) and El Niño events (during which warm surface water accumulates in the Pacific, heating the atmosphere). Such phenomena can for a time obscure temperature increases due to increased CO_2 concentrations. That is why a plot of global average temperature over recent decades—whether measured in the atmosphere, the oceans, or both—is a jagged line—though, on the whole, a rising one.

Still, if these other variables could be held constant while emission scenarios were varied, then global average temperature at any future time during the next few millennia would increase more or less continuously with higher cumulative carbon emissions scenarios up to that time. Of course, we can't do a planet-wide experiment to confirm that. But the principle is supported by atmospheric models and, ultimately, by basic physics. Its essence lies in the long-established fact that heat retention increases directly with atmospheric CO_2 content, together with the law of conservation of energy.

Even holding other variables constant, however, global average temperature is not a *perfectly* continuous function of cumulative CO_2 emissions. A certain microscopic graininess is introduced, for example, by the fact that the smallest unit of CO_2 is the molecule. But there are no *macro*scopic discontinuities.

There are, of course, tipping points in the climate system at which positive feedbacks accelerate the temperature increase. But these are variations in the steepness of the rising plot of temperature against cumulative CO_2 emissions, not discontinuities. Passing a climate threshold produces a quickening of temperature increase, not a discrete jump. Again, if we could hold other variables constant and vary only emissions scenarios, then for all practical purposes global average temperature during any given future period within the next few millennia would to a close approximation increase *continuously* (although not always at the same rate) with cumulative CO_2 emissions through the end of that period.

The degree to which global average temperature depends on carbon concentrations is only approximately known. But the equilibrium climate sensitivity (long-term global average surface warming following a doubling of CO_2 concentration relative to pre-industrial levels) is probably in the range of 1.5–4.5 °C (IPCC 2014, p. 62). A reasonable estimate, widely used in the literature, is 3°C. Assuming that a doubling of the pre-industrial CO_2 concentration raises the temperature 3°, raising it 6° requires four times that concentration—though this is only approximate. Each additional increment of CO_2 therefore produces a slightly smaller increase in temperature. We should not, however, assume that each additional increment produces a smaller increase in *harm*, for we have yet to examine how harm depends on temperature. That requires a definition of harm.

Harm Defined

By "harm" I mean bodily harm or death either to human or to nonhuman individuals. For humans, casualty or mortality rates are among the most useful objective, comprehensible, and ethically informative measures of harm (Nolt 2015b). A casualty rate is the number of people of a specified group who die or suffer injury or illness from a given cause over a specified period of time. (Sometimes people displaced and made homeless may also be included as casualties.) Mortality (fatality) rates are similar, but count only deaths, not injuries or illnesses. Casualty rates, being more inclusive, are obviously higher, but since most of the extant literature deals only with mortality rates, they are not as readily available.

One clear advantage of mortality rates is that they require relatively little interpretation. There are difficulties, of course, in determining the causes of death—especially with a cause as remote as global carbon emissions. Yet causes can be teased apart with a fair degree of confidence by statistical methods, as is done in the studies cited in the next section.

The predominant economic conceptions of harm, by contrast, require considerable interpretation. Harm is loss of welfare, and the standard neoclassical economic conception of welfare is preference satisfaction. Economists typically assume that preference satisfaction is higher when economic activity is more robust. So, for example, they often take the Gross Domestic Product (GDP) of a nation—sometimes with corrections for satisfaction or frustration of non-market preferences

—as a proxy for that nation's aggregate welfare. There are many reasons to doubt, first, of the identification of welfare with preference-satisfaction and, second, the adequacy of economic activity as a proxy for welfare (Nolt 2015a: 70–78, 2015b: 348–350), but these need not detain us here.

The crucial point for our purposes is that for long-term accounting economists apply a so-called "social discount rate" to estimates of future welfare. A social discount rate is an annual percentage reduction the value of anything (including human welfare) as a function of its distance into the future. In general, social discount rates are supposed to represent people's collective willingness to trade present costs or benefits for future costs or benefits—willingness to pay being regarded as a measure of expected preference satisfaction. Suppose, for example, that saving a life today is worth a million dollars. Then, assuming a 3% annual discount rate it is worth about $227.00 today to save a similar life 200 years from now. At a 5% discount rate it is worth about 60 cents.

The justification and purpose of social discount rates is widely disputed even among economists (Zhuang et al. 2007). These rates are standardly regarded as the sum of two components. One of these, the marginal utility of consumption, represents the degree of people's tendency when economic growth is expected to spend rather than save for future benefits. The other, which is called the pure time rate or utility discount rate, represents (independently of economic growth) people's collective preference for benefits now rather than later. The utility discount rate is especially difficult to justify empirically. In a comprehensive survey, Frederick, Loewenstein, and O'Donoghue conclude that

> … virtually every core and ancillary assumption of the DU [discounted utility] model has been called into question by empirical evidence collected in the past two decades. … While the DU model assumes that intertemporal preferences can be characterized by a single discount rate, the large empirical literature devoted to measuring discount rates has failed to establish any stable estimate (Frederick et al. 2002: 393).

Even if economists could accurately elicit an empirical utility discount rate, it would be irrelevant to the assessment of future welfare, for it aims to represent the importance of harms to *future* people as the degree to which they contravene the preferences of *present* people. The result is plainly inaccurate—and unjust. If, analogously, we were to measure harms to people of one race by the degree to which they contravened the preferences of people of another race, the inaccuracy and injustice would be palpable. Since time of birth is no less arbitrary a criterion of discrimination than race, it is likewise inaccurate and unjust to evaluate harms to future people by the degree to which they contravene our present preferences.

Given the subjectivity inherent in economic assessment of future values, it is not surprising that economists themselves differ widely on the value of the discount rate to be used in formulating policy. The Nordhaus-Stern controversy is a notorious illustration of this point. The argument is mainly over the pure time (discounted utility) rate. Nicholas Stern (2007, 2008) adopts a pure time rate of 0.1%, a figure that reflects only the possible loss of value due to human extinction. William Nordhaus (2006, 2007) uses a pure time rate of 3%, based instead on his estimate of collective time preferences. Using Stern's lower rate, massive and immediate spending to mitigate climate change is economically justified. Using Nordhaus's higher rate it is not. Thus by varying the essentially subjective pure time discount rate, one can rationalize whatever level of spending one would like.

Nearly all philosophers who have considered the question, beginning with Ramsey (1928), who introduced the standard formula for the social discount rate, have advocated a pure time discount rate of 0—which is to say, no pure time discount at all.

Moreover, while it is reasonable when thinking of economic costs to factor in the marginal utility of consumption, there is no defensible justification for doing so when considering the *moral* significance of bodily harm. Thus for bodily harm it is reasonable to assume an overall social discount rate of 0.

To thus refuse to discount harms with their distance from us in time is not to ignore the probabilities of future events. Social discount rates do not reduce probabilities; they reduce the value of future welfare. Probabilities remain, as always, essential to rational decision-making. A small probability of a given number of deaths, for example, is of proportionately less ethical concern than is the certainty of that number. Still, an $n\%$ chance of a hundred thousand casualties two centuries from now is the ethical equivalent of an $n\%$ chance of a hundred thousand casualties today—and this remains true when $n = 100$ (Nolt 2015a, Sect. 4.1.2).

For all of these reasons, undiscounted casualty or mortality projections, rationally inferred from solid empirical data, yield more straightforward, objective and reliable estimates of future harm than do the usual economic assessments. Casualty and mortality counts are familiar to everyone. The news media report them regularly for all sorts of calamities, and historians use them to assess the relative severities of disasters (battles, storms, earthquakes, plagues, etc.) across wide stretches of time. There is, moreover, some evidence that public health concerns (which is what casualty and fatality rates express) are more likely than other ways of conceiving future harms to elicit constructive public concern (Myers et al. 2012).

Ideally, perhaps, harms to *non*human life should also be gauged by casualty and mortality rates—one for each species, possibly weighted by some measure of the ecological importance of the species. But for most non-human species, such rates are unknown or poorly known, and probably would mean little to most people even if they were better known. There are exceptions, of course, for prominent megafauna. Many people know, for example, about declines in the populations of song birds or polar bears. But these comprise a miniscule fraction of all species.

A more practical and widely used, but still objective, proxy for harm to non-human life is biodiversity loss, understood in the form of global species extinction rates. Many people understand what extinction is and have some sense of what it means for a species (or many species) to be threatened or endangered. The relation of biodiversity loss to climate change is reasonably well documented over geological timescales, and conservation biologists monitor current extinctions—though, given the rapidity of change, they are hard-pressed to do so. Even so, a good bit or relevant information is available and can to some extent be extrapolated for predictive purposes.

In sum, both extinction rates and human casualty and mortality rates are practical, meaningful, and objective measures of the harms of climate change. But how do they depend on temperature?

Harm as a Function of Temperature

This section explains and defends the second premise of my central argument: that harm to both human and nonhuman life over a given period p during the next few millennia increases directly and more or less continuously with global average temperature during p.

Because the harms of anthropogenic climate change may continue for many tens of thousands—perhaps even hundreds of thousands—of years, no estimate of cumulative harm can be meaningful unless some such time period is specified. We could, I suppose, try to estimate the total harm over all those millennia, but any such "estimate" would only be a guess. Projections over the next few millennia are also be guesswork, but perhaps can be made with a touch more confidence. Even projections over the next few centuries are extremely dicey. For that reason I have made p variable. Choose whatever period—and hence degree of uncertainty—you feel comfortable with. But be aware that, whatever period you choose, harms probably continue to accumulate beyond it.

In allowing p to begin now, I am implicitly assuming that harm is already occurring. There is plenty of evidence for that assumption, especially in the case of harm to humans. Estimates of the current *annual* global mortality due to climate change all cluster in the hundreds of thousands. I am aware of estimates or projections from four different sources. The most recent estimate or projection from each is as follows:

- Global Humanitarian Forum (2009) estimates the current global mortality rate due to climate-change-induced malnutrition, diarrhea, malaria, dengue fever, storms and weather-related flooding at about 300,000.
- Development Assistance Research Associates (2012) estimates 400,000 annual deaths currently and 700,000 by 2030. It extrapolates from data for many specific factors, including drought, floods, landslides, storms, wildfires, diarrhea, heat and cold illnesses, hunger, malaria, and vector-borne meningitis.
- The World Health Organization (2014) forecasts 250,000 climate deaths per year in the period 2030–2050, due to extreme heat, coastal flooding, diarrhea, malaria, dengue, and undernutrition.
- Springman et al. (2016) estimate that there will be 500,000 deaths annually by 2050 merely from reduced food availability.

Each of these studies uses somewhat different data sources, each looks at different proximal causes of death, and all use statistical methods to distinguish for each proximal cause, the excess deaths attributable to climate change. Together they demonstrate that data and methods for providing rough casualty estimates and projections are already available. There is, moreover, reasonably close agreement among them, though they are not wholly independent of one another.

In some respects they are probably underestimates. None of them, for example, accounts for the disruption and violence that is likely to be induced by migration and competition for scarce resources as climate disruption worsens. It has been argued, for example, that drought induced by climate change helped to spark the Syrian civil war and the resultant migrations into Europe and elsewhere (Gleick 2014).

While climate change has some beneficial effects (e.g., more moderate temperatures in historically cold climates or carbon fertilization), as regards human health, it is already doing more harm than good (Kim et al. 2014). The ratio of harms to benefits will, moreover, inevitably increase as temperatures rise, magnifying heat waves, ice melt, sea-level rise, floods, droughts, the severity of storms, crop failures and resultant famine, and the spread of tropical diseases. Indeed (though my argument does not depend on this) harms are likely to increase more than proportionally relative to temperature. One reason is that the viable temperature ranges for populations of humans, livestock, crops, and other organisms on which humans depend (including those that supply ecosystem services) tend to exhibit a normal distribution—that is, a bell curve. Moving from the center toward the high-temperature end of such ranges produces little harm at first, but then, precipitously, accelerating harm, followed (if things get bad enough) by a decelerating decline to extinction. Another reason that harms are likely to increase more than proportionally relative to temperature is that sea level rise—and consequently coastal flooding and the severity of storm surges—are expected to accelerate with increasing temperature (Fasullo et al. 2016). Broome (2012: 33–36) provides additional reasons.

Admittedly, the current global casualty rates for climate change are fairly low relative to other important causes of death and injury. Mortality rates for anthropogenic air pollution, for example, are in the low millions annually (Silva et al. 2013), roughly an order of magnitude greater than the rates for climate change. Some might therefore suppose that the harms of climate change are relatively insignificant.

That would, however, be a serious mistake. CO_2 emissions cause harm in a way that has few parallels in ordinary human experience. Familiar harms are one-time effects of one-time causes or, at most, ongoing effects of ongoing causes. But the harms of CO_2 emissions are ongoing effects of one-time causes (emissions). Because these harms continue long after emissions cease, they can be very large in total. The harms of air pollution (particulates, ozone, sulfur dioxide, nitrogen oxides, etc.), by contrast, are generally local or regional, and once emissions cease, the pollutants are rapidly dispersed, diluted or degraded, so that health effects diminish rapidly. Even in the case of chlorofluorocarbons, which erode the Earth's ozone layer, harms diminish to negligibility within decades after emissions cease. But the harms of climate disruption continue for much longer than that. Thus, although anthropogenic CO_2 emissions will decline sharply within the next couple of centuries (we may hope, within the next couple of decades), the climate disruptions they cause may continue to inflict injury for as far as we can see into the future.

Consider, to be specific, just the period from now to 2100. Even assuming that current mortality rates remain steady (they are projected to increase), climate disruption will cause tens of millions of deaths by the end of this century (see Broome 2012, p. 33). This year's carbon emissions will contribute, however, not only to harms that occur during this century, but to harms occurring long after. It is not out of the question that the number of climate *casualties* over the next millennium, for example, will be in the billions (Nolt 2011a, 2013a).

Harm is, moreover, for all practical purposes a *smooth and continuous* function of temperature. In part the continuity results from the fact that, except when it takes the form of death, harm is almost always a matter of degree. Injuries, diseases and harms of deprivation can all vary continuously in severity. A little more heat in a drought, for example, incrementally decreases crop yields and water availability, which increases hunger, thirst and the health effects thereof. In part, too, this continuity is due to the vast number of individual harms (especially if we include harms to non-humans). Even if all harm came in discrete quanta, large-scale aggregate harm would still be a more or less continuous function of temperature.

It should be noted, incidentally, that not all the harms of CO_2 emissions are due to rising temperatures. The most significant exception is ocean acidification, which results from dissolution of atmospheric CO_2 into the seas. A recent statement on this phenomenon by the world's academies of science asserts that:

- The rapid increase in CO_2 emissions since the industrial revolution has increased the acidity of the world's oceans with potentially profound consequences for marine plants and animals especially those that require calcium carbonate to grow and survive, and other species that rely on these for food;
- At current emission rates models suggest that all coral reefs and polar ecosystems will be severely affected by 2050 or potentially even earlier;
- Marine food supplies are likely to be reduced with significant implications for food production and security in regions dependent on fish protein, and human health and wellbeing;
- Ocean acidification is irreversible on timescales of at least tens of thousands of years (IAP 2009).

Ocean acidification, too, increases essentially continuously with cumulative carbon emissions. It imperils marine life and hence humans (through reductions in global food supply).

What of the effects of temperature on nonhuman life? Here the picture is, if anything, clearer: rising global temperatures accelerate extinctions globally. Estimates compiled from a wide variety of studies and synthesized into a single plot represent the extinction rate as a smooth, continuous, and accelerating function of temperature (Urban 2015, Fig. 2). It is evident, then, that, other things being equal, species extinctions over a given period p during the next few millennia increase directly with global average temperature up through p. Species extinctions are, of course, discrete events, but we are taking them as a proxy for harm to nonhuman life, which varies continuously in just the way that harm to human life does. Hence, other things again being equal, harm to nonhuman life over a given period p during the next few millennia increases directly *and continuously* with global average temperature up through p.

Biodiversity losses, incidentally, are extremely long-lasting. Human populations can recover from losses within a generation or two, but biodiversity cannot—at least not by natural means. In examining the possibility that human activities (including greenhouse gas emissions) are causing Earth's sixth mass extinction, Barnosky et al. (2011) write that

> Recovery from today's biodiversity losses will not occur on any timeframe meaningful to people: evolution of new species typically takes at least hundreds of thousands of years and recovery from mass extinction episodes probably occurs on timescales encompassing millions of years (p. 51).

Humans might restore biodiversity artificially by genetic engineering, but it would not be wise to leave our descendants no choice but to venture the experiment.

Harm as a Function of Cumulative Carbon Emissions

To summarize: as carbon emission scenarios are varied, we see that, for any given period p over the next few millennia, it is to a close approximation true that

> Global average temperature during p increases directly and continuously with cumulative emissions between now and the end of p, and
> Harm to human and nonhuman life during p increases directly and continuously with global average temperature during p.

It follows that

> Harm to human and nonhuman life during p increases directly and continuously with cumulative anthropogenic carbon emissions from now through p.

I have considered just one independent variable: cumulative carbon emissions. Other variables, some of which we can (in theory, at least) control, will no doubt influence what actually happens. Adaptation or geo-engineering could diminish rates of harm, but may cause new harms (including international military conflict) of their own. Still, regardless of how other variables ultimately affect the outcome, each incremental increase in cumulative emissions can be expected to cause more harm.

The Non-identity Objection

Oddly, some theorists have doubted that we can harm distant future people. Steve Vanderheiden, for example, writes:

> Given the apparent impossibility of our present policy choices directly harming particular future persons, it appears that we cannot have any duties with respect to them, including negative duties not to harm them and positive duties to assist them, since neither is possible, at least insofar as these obligations are to persons (Vanderheiden 2008, p. 123).

The source of such this strange worry is the non-identity problem.

For our purposes, the problem can be expressed as a simple choice between two policies: greenhouse gas emissions as usual, on the one hand, and a steep and permanent reduction of emissions, on the other. Such big policy choices affect employment patterns, and ultimately reproductive choices. After a policy is changed, many young adults take different jobs, meet different mates, and have different children than they would have had otherwise. Their children in turn produce different children, and these differences ramify. Thus, after many generations, the resulting population may be composed of wholly different people from the one that would have resulted had the policy not been changed. So, regardless of which policy we choose, there will come a time in the distant future (say 2300) when none the people then alive would have been born had we made the other choice.

Now suppose we choose to pursue business as usual, and for centuries thereafter many millions of people are killed or injured by the resulting climate disruptions. But suppose also that for the most part, this does not make the lives of the people born after 2300 so bad that it would have been better for them never to have been born. Then it seems that we have not harmed these people. For an action can harm people only if it makes them worse off than they would have been otherwise; but had we chosen otherwise, these people would never have been born, and they are not worse off than that. Hence by pursuing business as usual, we apparently will not have harmed them. More generally, it seems to follow that we cannot harm distant future people. That was Vanderheiden's concern.

This puzzle has spawned an extensive literature. There are several disputed points. But, in brief [for more detailed accounts see Nolt (2015a, Sect. 4.5 or 2013a, pp. 114–116)] the crucial mistake lies in the assumption that an action can harm people only if it makes them worse off overall than they would have been otherwise. This assumption is false; for an action can harm a person simply by injuring her, even if it does not make her worse off on the whole than she would have been otherwise.

Imagine a woman living in 2300 who has a generally happy life but then is crippled in a hurricane whose extraordinary intensity is due to prior greenhouse gas emissions. The business-as-usual policy had two consequences for her: her being born and her becoming crippled. The former was a benefit, the latter a harm (in a quite ordinary sense of that term). The policy was not worse for her overall, since it made her life possible, and that was better for her than the alternative: that we had reduced greenhouse gas emissions so that she was never born. Nevertheless, our policy did make her worse off, and hence harmed her, by contributing to the hurricane. Therefore it is false that an action can harm people only by making them worse off overall than they would have been otherwise. We can in fact harm distant future people in just the way that we can harm present people—by, for example, contributing to storms that cripple them.

Conclusion: Significant Harms from Small Emissions

Harms consequent to anthropogenic CO_2 emissions will occur globally for centuries, perhaps millennia, perhaps even longer, after the fossil fuel era ends. It follows that cumulative harms will be very great. The total magnitude of these harms depends (to a close approximation) *continuously* on humanity's cumulative total carbon emissions. Thus each emission is harmful in proportion to its share of the total emissions.

Of course, no particular harm is attributable to the emissions of any specific individual, corporation, or nation. The CO_2 emitted by any of them is mixed in the atmosphere with the CO_2 from all other sources, and the harm results from the combination.

Still, the cumulative harm over centuries is so vast and so directly dependent on total CO_2 emissions that even small emissions cannot be assumed to have negligible effects. On the contrary, given the dependence of harm on cumulative emissions, it is reasonable to attribute to each emitter a fraction n/t of the total harms that anthropogenic emissions will produce over the coming centuries, where n is the mass of that person or organization's emissions and t is the mass of the cumulative total anthropogenic emissions. I have elsewhere estimated *very roughly* that the harm over the next millennium attributable to the lifetime carbon emissions of an average American is one or two future human casualties (Nolt 2011a, 2013a, b). Much greater harms are, of course, causally attributable to corporations or nations.

References

Allen M et al (2009) Warming caused by cumulative carbon emissions towards the trillionth tonne. Nature 458:1163–1166
Archer D et al (2009) Atmospheric lifetime of fossil fuel carbon dioxide. Annual review of earth and planetary sciences 37
Barnosky A et al (2011) Has the Earth's sixth mass extinction already arrived? Nature 471:51–57. https://doi.org/10.1038/nature0967
Broome J (2012) Climate matters: ethics in a warming world. W. W. Norton, New York
Development Assistance Research Associates (DARA) (2012) Climate vulnerability monitor, 2nd edn
Fasullo JT et al. (2016) Is the detection of accelerated sea level rise imminent? Sci Rep 6(31245), https://doi.org/10.1038/srep31245, http://www.nature.com/articles/srep31245
Frederick S, Loewenstein G, O'Donoghue T (2002) Time discounting and time preference: a critical review. J Econ Lit XL:351–401
Gleick PH (2014) Water, drought, climate change, and conflict in Syria. Weather Clim Soc 6:331–340. https://doi.org/10.1175/WCAS-D-13-00059.1
Global Humanitarian Forum (GHF) (2009) Climate change: the anatomy of a silent crisis. http://www.ghf-ge.org/human-impact-report.pdf
IAP (Interacademy Panel on International Issues) (2009) IAP statement on ocean acidification, http://www.interacademies.net/File.aspx?id=9075. Accessed 25 Jan 2014
Intergovernmental Panel on Climate Change (IPCC) (2014) Climate change 2014: synthesis report. Geneva, Switzerland

Kim K et al (2014) A review of the consequences of global climate change on human health. J Environ Sci Health, Part C 32(3):299–318. https://doi.org/10.1080/10590501.2014.941279

Matthews H, Solomon S (2013) Irreversible does not mean unavoidable. Science 340:438–439. https://doi.org/10.1126/science.1236372

Myers T, Nisbet M, Maibach E, Leiserowitz A (2012) A public health frame arouses hopeful emotions about climate change. Clim Change 113:1105–1112. https://doi.org/10.1007/s10584-012-0513-6

Nolt J (2011a) How harmful are the average American's greenhouse gas emissions? Ethics Policy Environ 14(1):3–10

Nolt J (2011b) Greenhouse gas emission and the domination of posterity. In: Arnold D (ed) The ethics of global climate change. Cambridge University Press, Cambridge, pp 60–76

Nolt J (2013a) Replies to critics of 'how harmful are the average American's greenhouse gas emissions?'. Ethics Policy Environ 16(1):111–119

Nolt J (2013b) The individual's obligation to relinquish unnecessary greenhouse-gas-producing devices. Philos Public Issues (New Series) 3(1):139–165

Nolt J (2015a) Environmental ethics for the long term. Routledge, London

Nolt J (2015b) Casualties as a moral measure of climate change. Clim Change 130(3):347–358. https://doi.org/10.1007/s10584-014-1131-2

Nordhaus W (2006) The Stern review on the economics of climate change. NBER Working Paper No. W12741. National Bureau of Economic Research, Cambridge

Nordhaus W (2007) A review of the Stern review on the economics of climate change. J Econ Lit 43(5):686–702

Ramsey F (1928) A mathematical theory of saving. Econ J 38:543–559

Shue, H (1993) Subsistence emissions and luxury emissions, Law & Policy 15(1):39–59; reprinted in Gardiner et al (2010), pp 200–214

Silva RA et al (2013) 'Global premature mortality due to anthropogenic outdoor air pollution and the contribution of past climate change. Environ Res Lett 8(3):1–11, http://iopscience.iop.org/1748-9326/8/3/034005/pdf/1748-9326_8_3_034005.pdf

Springman M et al (2016) Global and regional health effects of future food production under climate change: a modeling study. The Lancet. https://doi.org/10.1016/S0140-6736(15)01156-3

Stern N (2007) The economics of climate change: the Stern review. Cambridge University Press, Cambridge

Stern N (2008) The economics of climate change. Am Econ Rev 98(2):1–37; reprinted in Gardiner et al (2010), pp 39–76

Stocker T (2013) The closing door of climate targets. Science 339:280–282. https://doi.org/10.1126/science.1232468

Urban MC (2015) Accelerating extinction risk from climate change. Science 348(6234):571–573

Vanderheiden S (2008) Atmospheric justice: a political theory of climate change, Oxford

World Health Organization (WHO) (2014) Quantitative risk assessment of the effects of climate change on selected causes of death, 2030s and 2050s. http://www.who.int/globalchange/publications/quantitative-risk-assessment/en/

Zeebe R (2013) Time-dependent climate sensitivity and the legacy of anthropogenic greenhouse gas emissions. Proc Nat Acad Sci, 110. https://doi.org/10.1073/pnas.1222843110

Zhuang J, Liang Z, Lin T, De Guzman F (2007) Theory and practice in the choice of social discount rate for cost-benefit analysis: a survey. Asian Development Bank. http://www.adb.org/publications/theory-and-practice-choice-social-discount-rate-cost-benefit-analysis-survey

Chapter 4
Justice in Mitigation After Paris

Darrel Moellendorf

Abstract Justice between generations requires that the present generation takes significant steps to limit and then halt global warming. International justice requires that this be done in a manner that is consistent with poorer states continuing to pursue energy intensive, poverty eradicating human development strategies. The de-centralized process of pledging emissions reductions incorporated in Paris Agreement provides significant protection to poor states, and it is to be cheered by advocates of international justice. But this same process is thus far inadequate to the task of realizing intergenerational justice. States must increase the ambition of their pledges significantly. And the burden in that regard must fall primarily on wealthy countries in order to ensure that poverty eradicating human development can continue where it is needed. But collective action problems may undermine efforts to ramp up ambition. If the price of renewable energy does not fall sufficiently, states may be likely to shirk their responsibilities. Even if the price of fossil fuels continues to fall, the political influence of the fossil fuel industry could frustrate the mitigation effort. The best prospects for achieving justice in mitigation after the Paris Agreement lies in the success of movements that seek to redirect energy investment and policy towards renewable energy.

Introduction

Anthropogenic climate change already affects the lives and well-being of hundreds of millions of people and will do so into the foreseeable future. The change is driven by the use of greenhouse gases, most importantly CO_2. According to recent

---

D. Moellendorf (✉)
Cluster of Excellence, the Formation of Normative Orders, Institute of Political Sciences, Goethe-University Frankfurt, Lübeckerstraße, 60323 Frankfurt, Germany
e-mail: darrel.moellendorf@normativeorders.net

© Springer International Publishing AG, part of Springer Nature 2018
C. Murphy et al. (eds.), *Climate Change and Its Impacts*,
Climate Change Management, https://doi.org/10.1007/978-3-319-77544-9_4

scientific projections summarized by the Intergovernmental Panel on Climate Change (IPCC), the mean equilibrium surface temperature of the Earth by the end of this century is likely to be 2.6–4.8 °C higher unless we begin to reduce global emissions.[1] Warming at that rate is unprecedented in human history. Its effects are likely to be devastating, including widespread species loss and destruction of eco-systems, heat waves, extreme precipitation and tropical storms, and large and irreversible sea-level rise from terrestrial ice sheet melting. The rise in sea level could swamp low-lying island nations, such as Kiribati and Tuvalu, and could threaten major coastal cities including New York, Mumbai and Shanghai. Droughts could threaten food and water security in parts of Africa. Higher temperatures and increased precipitation in other area are expected to increase the risks of food- and water-borne as well as vector-borne diseases such as malaria. Effects such as these will create poverty traps in some already poor parts of the world, and are likely to slow economic growth globally. Droughts, sea-level rise, and flooding will encourage migration. Resource stresses and migration could increase international tensions and induce violent conflict.[2]

Mitigation policies seeking to reduce the extent of climate change by means of reducing emissions, preserving forests, and afforestation have the effect of redistributing some of the costs of climate change from the future to the present. In order that people in the future suffer fewer and less devastating effects, people in the present and near future must curb and then halt emissions and promote the health of forests. How burdens are distributed between generations is a matter of intra-generational justice. And insofar as climate change policy transfers costs to the present, it raises questions of how those costs should be distributed globally. Reducing global emissions will require international cooperation. Such cooperation will involve states assuming and effectively pursuing mitigation aims. How those aims are distributed among states is a matter of international justice. An international mitigation regime is likely to raise the costs of using fossil fuels as a means of reducing its consumption and of increasing the competitiveness of renewable energy. That could raise the absolute costs of energy around the world. But human development requires energy. One important question in this regard is whether some states have a claim against others to have their efforts at poverty eradicating human development protected from possible increases in absolute energy costs.

In the following two sections of this chapter I discuss considerations of intergenerational and international justice that seem appropriate to climate change policy. After that in the subsequent sections I discuss the prospect for achieving justice in these areas in the wake of the Paris Agreement of the United Nations Framework Convention on Climate Change (UNFCCC) in December of 2015.

---

[1] Intergovernmental Panel on Climate Change, *Climate Change 2014: Synthesis Report, Summary for Policy Makers*, pp. 10–11.
[2] Ibid., pp. 13–16.

Justice Between Generations

Climate change is driven primarily by the concentration of greenhouse gases in the atmosphere.[3] These concentrations raise the risks of various devastating events. CO_2 concentrations in the atmosphere have been built up by the emissions of generations going back to the beginning of the Industrial Revolution, but over half the stock of atmospheric CO_2 concentrations have been added from emissions in the last 60 years.[4] Climate change policy on behalf of future generations cannot only be conceived then as guided by a principle demanding that we rescue from a harm imposed by others. For our emissions are also part of the problem. Nor, however, is the moral requirement captured simply by a principle of refraining from causing harm since our response is required in part because of what others have done. Mitigation policy is important, then, in light of both risks inherited from previous generations and the risks that come to be as a result of our own actions. The principle not to compound already existing risks captures both aspects of the problem. Since compounding is a kind of causing, this principle is a specification of the duty not to cause undeserved suffering, albeit fine tuned to the greater risk of doing so because of the actions of others.

The application of the duty not to compound the risks of undeserved suffering is, however, complicated by the fact greenhouse gas production is an externality of the economic activity that also creates employment, infrastructure, and wealth; and all of these are not only enjoyed in the present but redound to future generations as well. Some children are not born into poverty because their parents escaped poverty by means of decent paying jobs in factories that emit greenhouse gases. And electricity powered by coal furnaces improves the health and educational outcomes of children. So, in considering the distributional effects of climate change mitigation policies the suffering that climate change risks is relevant, but so also are the benefits of the economic activities that are driving climate change. Our energy policies can distribute benefits and risks in various ways. A principle for guiding the appropriate distribution of those risks seems necessary to guide policy.

One prominent principle for distributing the risks and benefits of energy policy intergenerationally comes from the Economics literature. Discounted utilitarianism seeks to guide climate policy by the aim of optimizing consumption over an infinite time horizon. It is not, however, simply intergenerational consumption that is optimized, rather it is consumption subject to a discount rate that applies especially to future consumption. The related literature is full of debate about what the appropriate discount rate should be.[5] This debate is often fairly technical. For our purposes the technicalities can be avoided. Utilitarian approaches generally suffer

---

[3]Ibid., p. 6.
[4]Stocker et al. (2013).
[5]Much of the debate has centered around Nordhaus (2008) and Stern (2006). I discuss these in *The Moral Challenge of Dangerous Climate Change: Values, Poverty, and Policy* (Cambridge University Press, 2014).

from the problem of permitting massive burdens to be assigned to parties—in the present case to generations—for the sake of very minor gains to a sufficient number of other parties. An alternative distributive principle minimizes the difference in the ratio of global climate change costs to global GDP across generations. If burdens are thought of as costs in proportion to the ability to pay them, then the alternative principle approximates equal burdens across generations.[6] This approach does not employ a discount rate and therefore avoids the complications of appropriately setting one. The approach also claims the advantage of better satisfying our precautionary concerns.

Climate negotiators, however, under the influence of several states and civil society movements have eschewed adopting a particular intra-generational distributive principle in favor of adopting a limit on mean global warming. In 1996 the European Commission first adopted the goal of limiting warming to 2 °C.[7] At the meeting of the UN Framework Convention on Climate Change (UNFCCC) in Copenhagen in 2009 member parties affirmed the goal of limiting mean global warming to 2 °C above pre-industrial levels.[8] They then re-affirmed that goal the following year in Cancún.[9] In June of 2015 the G7 also adopted the goal. And at the Paris meeting of the UNFCCC parties agreed to limit warming to well below 2 °C.[10] Limiting warming to no more than 2 °C has become a major international policy goal. But is a 2 °C warming limit required by intergenerational justice? One reason to think that a precise temperature target cannot be a goal required by justice is that there is simply too much uncertainty involved in both our understanding of the climate system and the consequences of our policy. There is uncertainty about how to hit the goal since the atmospheric concentration of CO_2 that would limit warming to 2 °C is not known precisely. But there is also uncertainty about what the effects of limiting warming to 2 °C will be. How much seal-level rises depends on the dynamic collapse of terrestrial ice sheets. But that is too little understood to know at what temperature it will occur.

Still, there is good reason to think that the 2 °C goal is a good approximation of what intergenerational justice requires of climate policy. The risks of climate change accumulate as warming increase. Limiting warming to 2 °C is probably feasible, and can probably be achieved without laying extremely heavy costs on the present generation. The cost of producing energy by means of photovoltaic cells is dropping. That makes the 2 °C goal less expensive for present generations and therefore the burdens on the present generation of achieving it seem more reasonable. Alternatively one might argue that as our technology develops we could do more to adapt at a lower cost and therefore we should worry less about hitting an ambitious climate target such as the 2 °C one. Funding adaptation to climate change

---

[6]See Moellendorf and Schaffer (2016).
[7]European Commission Press release (2017).
[8]UNFCCC (2009).
[9]UNFCCC (2010).
[10]UNFCCC (2015).

is also a requirement of justice, but emphasizing adaptation at the expense of ambitious mitigation would be reckless with the well-being of future generations. There is an uncertain probability of catastrophic changes, such as rapid land-based ice sheet melting, and although we do not know the probability of such threats we can be reasonably confident that they increase as temperatures increase. And there is the additional worry that warming significantly beyond 2 °C could produce changes so severe that we could not properly adapt to them. Our capacity to adapt could be outstripped by the catastrophic effects of warming. There is some momentum coming out of the Paris Agreement to strive for an even lower goal of 1.5 °C, especially in light of the threat that climate change poses to low lying islands. Since the Earth has already warmed 0.8 °C, 1.5 °C would be a very ambitious goal. But if the costs of renewable energy continue to fall it might perhaps be attainable. The policy requirements of achieving it are currently being studied by the IPCC.

Part of the argument in favor of an ambitious warming limit invokes the uncertainty of catastrophic effects. Uncertain outcomes are distinguishable from low probability ones. Epistemic uncertainty about outcomes exists when the processes are not well enough understood to predict.[11] Sea-level rise due to the thermal expansion of the oceans and melting of glaciers is projected to be in the range of 0.52–0.98 m this century, unless we take additional action to mitigate climate change.[12] That's a significant rise of the sea-level in a short period of time. And it is likely to cause a lot of damage. According to one report 150 million people, three times more than now, would be exposed to a 1 in 100 year flooding event due to higher storm surges caused by that much sea-level rise.[13] It might, however, be much worse than that. There is epistemic uncertainty about the collapse of terrestrial ice sheets and the consequences of such collapse of sea level rise. The IPPCC's *Fifth Assessment Report* discusses the uncertainty as follows:

> There is high confidence that sustained warming greater than some threshold would lead to the near-complete loss of the Greenland ice sheet over a millennium or more, causing a global mean sea level rise of up to 7 m. Current estimates indicate that the threshold is greater than about 1 °C (low confidence) but less than about 4 °C (medium confidence) global mean warming with respect to pre-industrial [mean temperature]. Abrupt and irreversible ice loss from a potential instability of marine-based sectors of the Antarctic ice sheet in response to climate forcing is possible, but current evidence and understanding is insufficient to make a quantitative assessment.[14]

One possible, but uncertain consequence, of warming beyond 1.5 °C or 2 °C is to raise significantly the risk of greater and more rapid sea level rise.

---

[11]Knight (1921), Chap. 8. This is further discussed in my *The Moral Challenge to Dangerous Climate Change*, Chap. 3.
[12]*IPPC 2013*, WG1, SPM, p. 25.
[13]Nicholls et al. (2008).
[14]*IPCC 2013* WG1, SPM. p. 29.

Normative theorists are not unanimous about how to specify circumstances in which greater caution would be warranted in the face of uncertainty.[15] It seems clear that the appropriate response to uncertainty should depend considerably on the circumstances. When resources are scarce we may not want to expend them protecting against harms that are highly unlikely. It is possible that the planet could come under devastating attack from an intergalactic malign force. But such a possibility hardly warrants directing scarce resources to the design and construction of a defense system. By contrast, the uncertain threat of rapid and significant sea-level rise due to terrestrial ice sheet melting seems to warrant taking additional precaution, It is important then to develop a clearer understanding of when precaution is warranted.

I suggest that the following four conditions are jointly sufficient to warrant precautionary action in circumstances of uncertainty about an outcome[16]:

1. The harmful outcome, should it come to pass, would occur by means that are in at least general terms understood.
2. Several of the understood causal antecedents of the harmful outcome are in place.
3. The outcome is sufficiently harmful that the reasons for avoiding it are very strong.
4. In comparison to the reasons for avoiding the harm, the reasons for pursuing aims foreclosed by avoiding the harm are far less compelling.

It is not hard to think of examples that would illustrate these conditions with respect to harm to ourselves. If purchasing a relatively inexpensive sale item from a store would require that I park in a patrolled no parking zone in which cars illegally parked cars are sometimes towed, then it would be folly for me to do so. Better to forego the benefits of the sale than to risk the uncertain possibility of having the car towed. In this example the costs would, of course, fall entirely on me. It is case in which prudence seems to dictate avoiding the negative outcome, even if its probability is uncertain. There are good reasons to believe it could happen. The costs would be high. And the benefit gained by taking the chance is comparatively minor. When the harm would fall on other people, it is morality, not prudence, that dictates precaution. Sea-level rise due to terrestrial ice sheet collapse satisfies the above conditions. In general terms there is no mystery about why ice sheets would collapses. We have seen enough melting and breaking of ice to know that the processes that might cause the collapse exist. The additional harm caused by significantly more sea-level rise (seven times more) is something we have very good reason to avoid. And mitigation, at least within the 2 °C limit, can be accomplished at comparatively minor costs. These considerations reinforce the case for the urgency to mitigate on behalf of future people.

---

[15]For elaboration of the skeptical view see for example Sunstein (2005), pt. 1 and Posner (2004), Chap. 3.

[16]Three of these four conditions are discussed in more detail in *The Moral Challenge of Dangerous Climate Change*, Chap. 3. I have added condition 2 to the three discussed in the book.

The demands of justice not to compound the risks of undeserved suffering to people in the future and to avoid the uncertain possibility of the catastrophic effects of climate change provide the moral basis of the justification for a policy of mitigation. But energy creation, of which climate change is an externality, also creates many benefits both to present and future generations. So, any plausible climate change mitigation policy needs to be guided by an appreciation of both the risks and the benefits. In light of these considerations the goal of limiting warming to 2 °C seems like a reasonable one.

International Justice

Achieving any particular temperature target for mitigation will require the cessation of the use of fossil fuels. This is because the temperature increase is driven by the atmospheric concentration of greenhouse gases, especially CO_2; and CO_2 remains in the atmosphere for so long before it is recycled back to the Earth's surface that for practical purposes what matters in hitting a temperature target are cumulative emissions since the beginning of the Industrial Revolution.[17] For any temperature target assumed there is then a finite historical CO_2 emissions budget. Over half of the budget for limiting warming to 2 °C has been consumed already.[18] Indeed, unless global emissions begin falling the total budget will be consumed by 2037.[19] Mitigation policy, then, must aim ultimately at a carbon free global economy. The deadline is likely to come before the end of the present century. But currently global emissions are still increasing. So, the first step in any effective global mitigation policy has to be to decrease global emissions of fossil fuels almost immediately. Methods of doing that whether through reducing subsidies for fossil fuels, statutory emissions limits, carbon taxes, or emission trading schemes will have the effect of making fossil fuels more expensive.

Much of the world desperately needs expanded access to inexpensive energy. There are over 2 billion people who currently live in energy poverty, which is understood by the International Energy Agency to mean that they lack either access to electricity or modern cooking fuels.[20] About 1.4 billion of these people lack access to electricity. There is a strong correlation between the extent of energy poverty in a country and its lack of human development (which the UN Human Development Program measures in terms of per capita income, education, and health).[21] It is not entirely clear which direction the causation runs, but it is obvious

---

[17]AR5, SPM, pp. 9–10.
[18]Ibid.
[19]See http://www.trillionthtonne.org/.
[20]International Energy Agency, *World Energy Outlook 2011*.
[21]International Energy Agency (IEA) (2010).

that modern well-developed economies require electricity and fuel.[22] 145 million people die prematurely each year from the indoor pollution caused by burning biomass. This outstrips both malaria and tuberculosis.[23] The burdens of energy poverty fall heavily on women and children. The onerous work of gathering fuel is typically their job.[24] Lack of street lighting makes women more vulnerable to sexual assaults when gathering fuel. Children suffer disproportionally from the respiratory diseases caused by indoor pollution and their learning is constrained by lack of electricity. A significant constraint on mitigation policy, then, must be not to hinder existing efforts to expand access to energy. Increasing the costs of fossil fuels risks increasing the costs of energy generally.

The centrality of the use of energy to the success of poverty eradicating development efforts explains the inclusion of the right to sustainable development in the 1992 UNFCCC treaty. Article 3, paragraph 4 states that, "The parties have the right to, and should promote, sustainable development."[25] This is the treaty framework that governs all UN climate change negotiations. Pressure from least developed and developing countries who insist that their efforts to eradicate poverty not be hindered by a mitigation agreement continues to be strong and steady. In the agreement reached in Paris in 2015 seems to be reiterated: "This Agreement, in enhancing the implementation of the Convention, including its objective, aims to strengthen the global response to the threat of climate change, in the context of sustainable development and efforts to eradicate poverty."[26] This sentence can plausibly be read as re-asserting that the aim of permitting sustainable development is a constraint of mitigation policy.

In the UNFCCC negotiating context, where least developed and developing countries are eager to prevent the cost of energy being raised by mitigation policies to such an extent that policies serving the aim of poverty eradication would be undermined, the right to sustainable development is best interpreted as a claim that developing and least developed states have on industrialized ones that climate change policies not harm national efforts to pursue development. So interpreted, the right protects the development aims developing and least developed states from mitigation policies that might slow development by raising the absolute cost of energy. Respecting the right would seem to require either that poor states have the liberty to continue to use inexpensive fossil fuels over the short term and the corresponding obligation on the part of developed states to reduce their emissions enough to offset that use or a claim on developed states for financial assistance in

---

[22]Stern (2016).

[23]IEA, *Energy Poverty*, 13–14.

[24]Ibid.

[25]United Nations Framework Convention on Climate Change. See https://unfccc.int/resource/docs/convkp/conveng.pdf.

[26]UNFCCC, Conference of the Parties, Twenty-first session (Paris Agreement), Article 2, paragraph 1. See https://unfccc.int/resource/docs/2015/cop21/eng/l09.pdf.

making the transition to more expensive renewable energy in a manner that does not slow development.

There are two justifications of the right to sustainable development that both invoke non-controversial and widely shared moral principles. The first justification rests on the principle of promise keeping. The parties who signed original 1992 United Nations Framework Convention on Climate Change treaty pledged to respect the right to sustainable development; such a pledge creates a promissory obligation to do so. To proffer a mitigation proposal that would predictably retard the human development of poor countries is a violation of the promissory obligation that states incurred in ratifying the 1992 treaty.

Even if states had not pledged to honor the right to sustainable development, they would nonetheless be duty-bound to respect the right since fairness would require it. In a cooperative enterprise, such as the international effort to mitigate climate change, parties should not be assigned a devastatingly heavy burden when that can be avoided by laying a comparatively much lighter burden on other parties. This is especially the case if the burden would include prohibiting the party from pursuing an aim it is morally required to pursue.

Mitigation policies that would hinder or slow poverty eradicating human development would assign poor states that kind of burden. Moreover, developing and least developing states are morally required to pursue poverty eradication in their countries. Fairness requires that developed states that are more able to bear the burden of mitigation, either by making deeper emissions cuts or by subsidizing renewable energy in poor countries, take on a heavier burden in order to prevent slowing poverty eradicating human development in poor countries.

The Paris Agreement

The agreement reached at the 21st Conference of the Parties (COP 21) of the UNFCCC in Paris is significant for several reasons, but perhaps most all because it seems to have salvaged an international negotiating process that had been in doubt ever since the COP 15 in Copenhagen in 2009 failed to produce the robust treaty that had been hoped for at the time. Agreement in Paris was possible in part because the process of making emissions reduction pledges was decentralized. The pledges made by states were not the product of diplomatic wrangling, as was the case with the Kyoto Protocol, but instead the result of states deciding on their own, through their own political processes, what they were willing to do. This policy process of making pledges to reduce emissions and later subjecting progress made on those pledges to international scrutiny is often referred to as "pledge and review."

The pledge and review process that gave rise to the Paris Agreement was, however, a mixed blessing. The good in the process is twofold. First, decentralization fostered broad agreement. Additionally, because states were to be subject only to obligations that they authored themselves, no state had a reason to find the burdens allotted by the agreement unreasonable and to withhold assent on that basis. Decentralization provided a procedural safeguard for the substantive norm of the right to sustainable development. Since states were not pressured into their obligations, they could be assured that obligation assumed was not going to constrain unduly their development objectives. By the lights of international justice, then, the Paris Agreement looks positive.

There is, however, plenty of reason for concern about the Paris Agreement. Most obviously the problem with a decentralized pledge process is that it does not ensure that the sum of the pledges is sufficient to hit the stated goal of limiting warming to well below 2 °C. Independent analyses of the pledges made in Paris provide reasons to think the total emissions that they allow would considerably overshoot the temperature goal. One recent report projects that the warming that would occur if the pledges were honored would be in the range of 2.6–3.1 °C.[27] At the high end, that is more than double the 1.5 °C that is also mentioned in the agreement as a desired goal.

Greater mitigation ambition is needed by states if the global warming limits affirmed in Paris are to have a good chance of being met. The need for states to increase their mitigation ambition is well recognized in the Paris Agreement. Article 4, paragraph. 3 of the Agreement states that, "Each Party's successive nationally determined contribution will represent a progression beyond the Party's then current nationally determined contribution and reflect its highest possible ambition."[28] So subsequent pledges by states are expected, and they are expected to be more ambitious than the state's previous pledge. In accordance with the review part of pledge and review, the agreement envisions a survey of progress in achieving the aims of the agreement (called a "stocktake") occurring at 5 year intervals beginning in 2023 (Article 14, paragraph 2). After each review, pledges would be expected to be renewed and increased. Due to the limit on cumulative emissions, if warming is to be limited to 2 °C, delaying the first stocktake till 2023 will require that subsequent pledges be very ambitious since from the time of making their initial pledge in 2015 states (even assuming they act in accordance with their pledges) would have been pursuing policies that overshoot the warming goal. Greater ambition sooner, of course, would require a less drastic change of course later.

One important question is whether states can be expected to keep the pledges that they have made. If renewable energy were to remain significantly more expensive than fossil fuels over the period until the first stocktake, then the effort to mitigate might be undermined by a collective action problem. Although each state has an interest in having warming limited so as to reduce the risks of climate change, it could be the case that no state has an economic interest in assuming the costs necessary to keep its pledge. If the costs of mitigation were to reduce economic growth somewhat or require businesses to assume costs without which they would be more competitive economically, then, if a sufficient number of other states

---

[27]Rogelj et al. (2016).
[28]Paris Agreement.

pursue mitigation policies it could be advantageous for a state not to do so since the mitigation goal would be in any case be approximated by the actions of other states, as long the state shirking is not one of the largest emitters. And, if other states do *not* pursue mitigation policies, then a state it would also be advantageous for a state not to mitigate since it would suffer a competitive disadvantaged by doing so.

In light of the danger that many states might reason along the lines just summarized, there is need for transparency in reporting activities so that it would be discernable when states are shirking. The Paris Agreement takes note of that need. It says in Article 4, paragraph 3 that "Parties shall account for their nationally determined contributions. In accounting…Parties shall promote environmental integrity, transparency, accuracy, completeness, comparability and consistency, and ensure the avoidance of double counting." Of course it is one thing to state a norm of transparency; it is another to realize it in practice. If realized, however, states would have some additional incentive to honor their pledges. If states are seen to be making good on their pledges, they will accrue reputational gains that might recompense some of their economic loss. Since mitigation is a long term cooperative project and there will be more than one stocktake, seeing other states make good on their proposals can foster trust and help to reduce the tendency to seek competitive advantage. That serves the aim of increasing the ambition of the pledges in the next round.[29]

There is also need for an authoritative body that could promote compliance with the pledges. In Article 15, paragraphs 1 and 2 the Agreement declares that "A mechanism to facilitate implementation of and promote compliance with the provisions of this Agreement is hereby established. The mechanism…shall consist of a committee that shall be expert-based and facilitative in nature and function in a manner that is transparent, non-adversarial and non-punitive." The inclusion of a Compliance and Implementation Committee is a nod to the problem of the incentives not to comply that could undermine the success of the agreement. It is noteworthy, however, that the committee should be non-adversarial and non-punitive. That is indicative of a fundamental problem in devising an institution that would encourage compliance. Widespread compliance is good, but each state would also like to avoid sanctions. So, although a strict system of punishment might go a long way towards undermining the collective action problem, it might also discourage states from entering into the agreement. Eleanor Ostrom refers to the problems of establishing enforcement mechanisms as a second order collective problem. The primary problem of keeping the agreement can only be solved by assurance that parties will comply.[30] That assurance is fostered by enforcement of the agreement, but establishing the incentive to enforce presents its own problems.

---

[29]The importance of iteration in building trust so as to prevent collective action problems is well appreciated. Elinor Ostrom discusses the issue *Governing the Commons: The Evolution of Institutions for Collective Action* (Cambridge: Cambridge University Press, 2015), especially Chap. 6.

[30]Ibid., Chap. 2.

In the Paris Agreement the representation structure of the Compliance and Implementation Committee has been established, but its membership has not yet.[31] It remains to be seen whether a non-punitive compliance mechanism will be effective in eroding the collective action problem.

In the social sciences climate change negotiations are often viewed as version of a Tragedy of Commons collective action problem. Ensuring commitment to an agreement is difficult because although all parties have an interest in the agreement producing its intended positive effects, many parties may lack sufficient interest in acting as the agreement demands no matter all the other parties do. Negotiations aim to build trust between through repeated rounds of give and take in which agreements are honored. The problem appears even more vexing, however, if the parties have high discount rates vis-à-vis the interests of future persons. Then the problem looks more like what the philosopher Stephen M. Gardiner calls the Intergenerational Storm[32]: Each generation has an interest in the previous reducing its emissions, but not in assuming the costs so as to reduce emissions on behalf of subsequent ones. This sort of problem is more vexing insofar as give and take between a previous and subsequent generation is impossible.

I shall not try to determine here whether the Tragedy of the Commons or The Intergenerational Storm has greater power to explain the difficulties of securing a lasting and effective agreement to mitigate climate change. I'll note only three things. First, the Tragedy of the Commons seems to have the capacity to focus on the appropriate agents, namely states, whereas the Intergenerational Storms seeks an explanation of generational behavior. Second, however, the Intergenerational Storm does, have a plausible explanation of why the development of a compliance mechanism is difficult. All state leaders, sharing a common generational membership, may have no interest in enforcing compliance. And, third, if the discount rate of state representatives is high in the Tragedy of Commons, the two explanatory frameworks do not seem to be significantly different. This points to the most important thing for present purposes, namely what the two frameworks share. Both frameworks take the problem as driven by parties' lack of interest in assuming the costs of climate change mitigation.

The Paris Agreement constitutes a step forward for international negotiations to mitigate climate change insofar as it has the potential to serve as the basis for broad international cooperation. That cooperation is necessary to serve the aim of intergenerational justice to minimize contribution to and prevent the risks and uncertainties of harm due to climate change. Broad cooperation is fostered by the decentralized means by which states commit to mitigating climate change. The decentralized process also helps to secure the aim of international justice requiring that the right to sustainable development be respected since states cannot be pressured into making mitigation commitments that would harm their pursuit of poverty eradicating human development. The principal worry regarding mitigation

---

[31]Voigt (2015).

[32]Gardiner (2001). The idea is introduced on pages 32–40.

Prospects for Progress

The collective action problem that threatens international cooperation is driven by the costs of transitioning from fossil fuels to renewable energy. As long as fossil fuels are cheaper, states have an interest in there being generalized mitigation, but no state may have an economic interest in assuming the costs of mitigation regardless of what other states do. But the severity of the collective action problem decreases as the cost of renewable energy in comparison to fossil fuels falls.

One reason that the comparative cost of renewable energy is falling is that the absolute cost of renewable energy, solar power and wind in particular, is dropping quickly. Although the levelized cost of coal is often estimated to be less than solar, the costs of solar are steadily falling.[33] Solar energy's share of the global market has doubled seven times in the last 15 years. Economies of scale are driving down costs; every time solar's share of the market doubles, costs fall 24%.[34] In contrast to solar's growth, coal consumption is dropping in OECD countries; and coal consumption also seems to be flattening out in China.[35] As solar's price falls, the gap between the comparative prince of solar and coal narrows.

Coal is also more expensive than often has been appreciated. The market price of coal does not fully incorporate its costs, which include environmental and health costs. A recent report by the International Monetary Fund argues that these costs amount to nearly 4 percent of global GDP.[36] In 2016 for the first time the International Energy Agency dedicated a *World Energy Outlook Special Report* to energy, air pollution, and health. The report finds that fossil fuel combustion in energy plants and industrial facilities is responsible for 3 million premature deaths each year.[37] As the real price of fossil fuels is understood to be higher than previously appreciated, the gap between the costs of renewables and fossil fuel may be seen to be less wide than previously calculated.

The gap between the comparative costs of some renewables and fossil fuels is closing because of a double movement of prices. Solar is becoming cheaper, and fossil fuels are more expensive than we have reckoned. Some studies now indicate

---

[33] International Energy Agency (IEA) (2015).
[34] Randall (2016).
[35] Ibid.
[36] Coady et al. (2015).
[37] OECD and IEA, *Energy and Air Pollution: World Energy Outlook Special Report 2016, Executive Summary*, p. 1.

that solar and wind are even cheaper than coal and gas in certain markets.[38] Insofar as a collective action problem driven by the cost of transitioning to renewable energy threatens to undermine the mitigation agreement, the closing of the gap between the cost of renewable energy and fossil fuels is very good news. And new mitigation policies can be expected to raise the price of fossil fuels. So, the competiveness of renewables should strengthen. The bad news is that that is not necessarily sufficient to remove the collection action problem. This is because action is guided by beliefs rather than facts. Until investors are confronted with irrefutable evidence and consumers with lower prices, it is possible for disinformation campaigns to remain effective.

Depending on the modelling assumptions made, studies indicate that one half to two thirds of all remaining fossil fuels reserves cannot be exploited if we are to have a reasonable chance of limiting warming to 2 °C.[39] That gives the fossil fuel industry a tremendous incentive to discourage mitigations efforts and falsify climate science. One way to do that, which has proven effective in other cases, such as tobacco, is to finance disinformation campaigns.[40] These campaigns can affect public opinion and therefore indirectly affect legislation. And where political systems allow fossil fuel companies to exercise political influence on legislation directly, we can expect them to do so. According to one study the fossil fuel industry spent almost $351 million donating to, and influencing, the 113th Congress of the US. That industry also received nearly $42 billion in federal production and exploration subsidies.[41] Clearly the donations were well spent. The influence that money buys can be a significant hindrance to justice in mitigation policy.

Even if the factual basis for the collective action problem is vanishing, the problem will not necessarily simply go away. Reasonable hope of limiting warming to 2 °C will require not only that the costs line up in favor of renewable energy but also that public opinion and political will do so as well. Achieving that will require public education and political struggle in many countries. Under the Trump administration, the USA is likely to be an especially important site of these efforts. There are efforts such as this underway in several countries. On many university campuses there are efforts to have the universities divest their portfolios from fossil fuels; and churches are also following suit.[42] There are campaigns to inform shareholders about the bad investment that fossil fuels are over the longer term.[43]

---

[38]Carbon Tracker Initiative (2016).

[39]Various studies are compared in IEA, "Can CO_2 Capture and Storage Unlock 'Unburnable Carbon'?" May 2016. Available on line at http://www.ieaghg.org/exco_docs/2016-05.pdf. (Accessed Sept. 24, 2016.)

[40]See Oreskes and Conway (2010), Chap. 6.

[41]Oil Change International (2016).

[42]See Fossil Free: http://gofossilfree.org/what-is-fossil-fuel-divestment/. (Accessed Sept. 24, 2016.)

[43]See Carbon Tracker Initiative: http://www.carbontracker.org/. (Accessed Sept. 24, 2016.)

And there are very important efforts to halt the construction of, and even shut down existing, coal fired power plants.[44]

Concluding Remarks

Justice between generations requires that the present generation takes steps to limit global warming. International justice requires that this be done in a manner that is consistent with poorer states continuing to pursue energy intensive, poverty eradicating human development strategies. The Paris Agreement lays out the basis for international cooperation to achieve justice in climate change mitigation. But much needs to be done. States must keep the pledges that they have made; and they must increase the ambition of their pledges significantly. The burden in that regard must fall first on wealthy countries in order to ensure that poverty eradicating human development can continue where it is needed. If the price of renewable energy does not fall sufficiently, states may be likely to shirk their responsibilities. Even if the price of fossil fuels continues to fall, the political influence of the fossil fuel industry could frustrate the mitigation effort. In either case, the best prospects for achieving justice in mitigation after the Paris Agreement lies in the success of movements that seek to redirect energy investment and policy towards renewable energy. Nothing less than intergenerational justice is riding on their success.

References

Carbon Tracker Initiative (2016) The end of the load for coal and gas. Available on line at http://www.carbontracker.org/report/the-end-of-the-load-for-coal-and-gas/. Accessed 24 Sept 2016
Coady D et al (2015) IMF working paper: how large are global energy subsidies? Available on line at https://www.imf.org/external/pubs/ft/wp/2015/wp15105.pdf. Accessed 23 Sept 2016
European Commission Press release (2017) Available on line at http://europa.eu/rapid/press-release_PRES-96-188_en.htm?locale=en. Accessed 4 Jan 2017
Gardiner SM (2001) A perfect moral storm: the ethical tragedy of climate change. Oxford University Press, Oxford
International Energy Agency (IEA) (2010) Energy poverty: how to make modern energy access universal. OECD/IEA, Paris, p 32
International Energy Agency (IEA) (2015) Nuclear Energy Agency, and Organization for Economic Cooperation and Development (OECD), The Projected Costs of Generating Energy 2015 Edition. Available on line at https://www.oecd-nea.org/ndd/pubs/2015/7279-proj-costs-electricity-2015-es.pdf. Accessed 23 Sept 2016
Knight FH (1921) Risk, Uncertainty, and Profit. Hart, Schaffner and Marx, New York
Moellendorf D, Schaffer (2016) Equalizing the costs of climate change. Midwest Stud Philos XL:43–62

---

[44]See the Sierra Club Beyond Coal: http://content.sierraclub.org/coal/. (Accessed Sept. 24, 2016.)

Nicholls RJ et al (2008) Ranking port cities with high exposure and vulnerability to climate extremes: exposure estimates, OECD Environment Working Papers, No. 1, OECD Publishing. http://dx.doi.org/10.1787/011766488208

Nordhaus W (2008) A question of balance: weighing the options on global warming policies. Yale University Press, New Haven

Oil Change International (2016) Fossil fuel funding to congress: industry influence in the US. Available on line at http://priceofoil.org/fossil-fuel-industry-influence-in-the-u-s/. Accessed 24 Sept 2016

Oreskes N, Conway EM (2010) Merchants of doubt: how a handful of scientists obscured the truth from tobacco smoke to global warming. Bloomsbury, New York

Posner R (2004) Catastrophe: risk and response. Oxford University Press, Oxford

Randall T (2016) Wind and solar are crushing fossil fuels. Bloomberg. April 6, 2016. Available on line at http://www.bloomberg.com/news/articles/2016-04-06/wind-and-solar-are-crushing-fossil-fuels. Accessed 23 Sept 2016

Rogelj J et al (2016) Paris Agreement climate proposals need a boost to keep warming well below 2 °C. Nature 534:631–639

Stern N (2006) Stern review: the economics of climate change. Cambridge University Press, Cambridge

Stern DI (2016) The role of energy in economic growth. In: Guruswamy L (ed) International energy and poverty. Routledge, New York, pp 11–23

Stocker TF et al (2013) IPCC, 2013: Summary for Policymakers (SPM). In: Climate Change 2013: the physical science basis. Contribution of Working Group I (WG1) to the Fifth Assessment Report of the Intergovernmental Panel on Climate Change. Cambridge University Press, Cambridge, p 12

Sunstein C (2005) Laws of fear: beyond the precautionary principle. Cambridge University Press, Cambridge

UNFCCC (2009) Report of the conference of the parties on its fifteenth session, held in Copenhagen from 7 to 19 December 2009. Available on line at http://unfccc.int/resource/docs/2009/cop15/eng/11a01.pdf. Accessed 13 May 2015

UNFCCC (2010) Report of the conference of the parties on its sixteenth session, held in Cancun from 29 November to 10 December 2010. Available on line at http://unfccc.int/resource/docs/2010/cop16/eng/07a01.pdf#page=2. Accessed 14 May 2015

UNFCCC (2015) Draft Decision CP 21. Available on line at http://unfccc.int/resource/docs/2015/cop21/eng/l09r01.pdf

Voigt C (2015) The compliance and implementation mechanism of the Paris agreement. Rev Eur Commun Int Environ Law 25:161–173

Chapter 5
Utilitarianism, Prioritarianism, and Climate Change: A Brief Introduction

Matthew D. Adler

Abstract This chapter compares prioritarianism and utilitarianism as frameworks for evaluating climate policies. Prioritarianism is an ethical view that gives greater weight to well-being changes affecting worse-off individuals. This view has been much discussed in recent moral philosophy but, thus far, has played little role in scholarship on climate change—where the utilitarian approach has, to date, been dominant. Prioritarianism and utilitarianism can be operationalized as policy-evaluation methodologies using the formalism of the "social welfare function" (SWF). Outcomes are converted into vectors (lists) of well-being numbers, one for each person in the population of concern. These lists are then ranked using some rule. The dominant approach in climate economics is to employ a discounted-utilitarian SWF. Well-being numbers are multiplied by a discount factor that decreases with time; these discounted numbers are then summed. The discounted-utilitarian SWF is problematic, both in incorporating an arbitrary preference for earlier generations, and in ignoring the well-being levels of individuals affected by policies. By contrast, the non-discounted prioritarian SWF eschews a discount factor, and adjusts well-being numbers so as to give priority to the worse off. This chapter describes the discounted-utilitarian and nondiscounted-prioritarian SWFs, and compares them with reference to three important topics in climate policy: the Ramsey formula, the social cost of carbon, and optimal mitigation.

Introduction

The "social welfare function" (SWF) is a standard tool in welfare economics (Adler 2012; Blackorby et al. 2005; Boadway and Bruce 1984; Kaplow 2008; Weymark 2016). Any given person, in a particular outcome, can be characterized as having a bundle of welfare-relevant attributes—attributes such as material consumption, health, leisure, psychological state, environmental quality, etc. This attribute bundle

M. D. Adler (✉)
Duke Law School, 210 Science Drive, Durham, NC 27708, USA
e-mail: adler@law.duke.edu

is converted into a well-being number. Each possible outcome (either the status quo outcome, or the outcome that would occur as a result of some policy intervention) becomes a vector (list) of well-being numbers—one for each person in the population of interest. An SWF is a rule for ranking these well-being vectors. It enables the policy analyst to compare outcomes in light of the patterns of well-being to which the outcomes correspond.

The SWF approach is widely used in climate economics (Botzen and van den Bergh 2014). For example, it was the basis both for the *Stern Review* (a major report for the U.K. government overseen by Nicholas Stern) and for William Nordhaus' well-known critical response to the *Stern Review* (Stern 2007; Nordhaus 2007, 2008).

The dominant functional form for the SWF, in climate economics, is *discounted utilitarianism* (Botzen and van den Bergh 2014). Individual well-being numbers are multiplied by a discount factor that is decreasing with time—so called "pure time preference." Later generations are thereby given less weight in social assessment than earlier ones. These discounted well-being numbers are then summed.

Despite its wide application, discounted utilitarianism is problematic—or so it can plausibly be argued. First, the use of a time-preference factor is ethically arbitrary (Ramsey 1928; Broome 2008; see also Dasgupta (2012) and Arrow et al. (2014), citing other scholars who have opposed time discounting). Second, even if the time-preference factor is removed, utilitarianism can be criticized for simply adding up well-being numbers [Much of contemporary moral philosophy is non-utilitarian, most famously Rawls (1999)]. Utilitarianism takes no account of the distribution of well-being.

These two criticisms of the discounted-utilitarian SWF can be answered by shifting to a different SWF: the non-discounted *prioritarian* SWF (Adler 2012, Chap. 5). Well-being numbers are adjusted via a concave transformation function.[1] By summing concavely transformed well-being, the non-discounted prioritarian SWF gives priority to well-being improvements (or the avoidance of losses) affecting individuals at lower well-being levels. It is thereby sensitive to the distribution of well-being. Moreover, because no time-preference factor is incorporated in the SWF's formula, there is no bias in favor of earlier generations.

Prioritarianism is a well-developed concept in moral philosophy, and has made inroads into welfare economics (Blackorby et al. 2005; Boadway and Bruce 1984; Brown 2005; Holtug 2010, 2015; Parfit 1991, 2012; Porter 2012; Tungodden 2003; Weymark 2016; Williams 2012). But it has had little influence, to date, on climate economics. In what follows, I set forth the discounted-utilitarian (DU) SWF and non-discounted prioritarian (NP) SWF, and then compare them with respect to climate policy—focusing specifically on the Ramsey formula, the social cost of carbon, and optimal mitigation.

---

[1]To be sure, this transformation function needs to be specified. See below for a discussion of how to do so.

The discussion is intended as a brief introduction to the topics covered. It draws in part from more detailed and formal treatments elsewhere [specifically Adler and Treich (2015); and Adler et al. (2017)], to which the interested reader is referred.

Welfare economics generally focuses on human well-being—and that is true, in particular, of the two SWFs considered here. No intrinsic ethical weight is given to the well-being of non-human animals, to the preservation of species, or to other environmental values. These matter, but only instrumentally—only to the extent they affect the welfare of human beings. Whether and how to move beyond this human-centered ethical perspective is a vital matter for debate in environmental ethics, but not one I will grapple with here.[2]

The Discounted Utilitarian SWF

In setting forth the formula for the DU SWF and below, for the NP SWF, I'll make some simplifying assumptions. Each of the simplifying assumptions is adopted, to a substantial extent, in climate policy modelling; there is nothing idiosyncratic in my use of them.

Time is divided into discrete periods and is finite. The first time period $t = 1$ is the present, and T_{max} denotes the final period.[3] The world is divided into regions. These are indicated by the variable "r," and are numbered from 1 to R, with R the total number of regions.

The population of each region at a given time is fixed. P_{tr} denotes the population of region r at time t. In reality, of course, the future population of the world is not fixed. It may change, both because of climate change, and because of other factors. Variable population poses thorny philosophical questions (Blackorby et al. 2005). See Adler (2009) on prioritarianism and variable population; and Adler and Treich (2015) with reference to climate change.

Variation in well-being among individuals within a given region is ignored. Moreover, it is assumed that well-being can be defined as a function of individual *consumption*. A person's "consumption," as that term is used by economists, means

---

[2]On the possibility of extending prioritarianism to animals, see Holtug (2007).

[3]Although some climate analysis assumes infinite time, it seems exceedingly unlikely that the earth or human species will continue ad infinitum. See Adler and Treich (2015); Adler (2012, pp. 576–79). Thus I assume that all the outcomes under consideration are such that the human species becomes extinct after a finite time T_{max}. To be sure, we don't know what T_{max} is! Such uncertainty, like other sources of uncertainty relevant to climate policy, can be handled by seeing each policy choice as a probability distribution over outcomes. In this instance, outcomes would differ in T_{max}, and the assignment of probabilities to outcomes would reflect our assessment of the likelihood of different such maximum times. The probability of a given T_{max} could be exogenous to climate policy or—even more realistically—affected by climate policy itself.

In this chapter, to keep the presentation simple, I generally ignore uncertainty—but it should be stressed that the SWF framework certainly has the resources to take account of uncertainty. See below, briefly discussing SWFs under uncertainty, and citing sources.

the total money value, at market prices, of the goods and services that the individual uses ("consumes"). Let $c_{tr}(x)$ denote the per capita consumption of individuals in region r at time t, in a given outcome x. Then (we are assuming) there is a well-being function $u(.)$ such that: the well-being of each person in region r at time t equals $u(c_{tr}(x))$.

But what about the non-market sources of well-being, such as an individual's health, or the quality of her environment? We can take account of non-consumption attributes either by (1) switching to a more complicated well-being function $v(c_{tr}(x), \mathbf{b}_{tr}(x))$, with $\mathbf{b}_{tr}(x)$ representing non-consumption attributes, or (2) retaining the simpler form $u(c_{tr}(x))$, but now letting c_{tr} represent normalized consumption—a quantity that starts with an individual's actual consumption, and then adjusts it up or down to reflect her health, environmental quality, etc. I'll follow course (2), and will henceforth use "consumption" to mean normalized consumption.[4]

A different question concerns the derivation of the well-being function $u(.)$. Where does this come from? Assume individuals in all regions and at all times have a common preference structure with respect to attribute bundles (bundles of both consumption and non-consumption attributes). This common preference structure consists in a ranking of attribute bundles and lotteries over attributes bundles. If the preference structure is well-behaved, it can be represented by a so-called "utility" function.[5] And the well-being function $u(.)$ is straightforwardly derived from the utility function that represents the common preference structure.

The assumption of common preferences can be relaxed, but this makes modelling more complex—since now the well-being of individuals in region r at time t is described as a function both of their (normalized) consumption *and* of the specific preference structure held by these individuals. I won't pursue this more complicated possibility here. Indeed, preference heterogeneity is typically ignored by climate analysts [On the derivation of a well-being function from preferences,

---

[4]This is conceptually straightforward. Let $v(.) = v(c, \mathbf{b})$ be a well-being function defined in terms of both consumption and non-consumption attributes. Arbitrarily choose some specific level $\mathbf{b}^+$ of the non-consumption attributes. Define $u(c)$ as $v(c, \mathbf{b}^+)$. Now, for a given bundle $(c, \mathbf{b})$, define c^{norm} as follows: c^{norm} is such that $v(c^{norm}, \mathbf{b}^+) = v(c, \mathbf{b})$. Note now that for a given bundle $(c, \mathbf{b})$, $u(c^{norm}) = v(c, \mathbf{b})$. So well-being comparisons in terms of $u(.)$ applied to normalized consumption perfectly mirror such comparisons in terms of $v(.)$ applied to consumption-nonconsumption bundles.

What is the advantage of expressing the analysis in terms of normalized consumption, rather than simply bundles of consumption and non-consumption attributes? Perhaps none, if we were starting from scratch. However, much existing work in economics (including climate economics) employs utility functions of the form $u(c)$ rather than $v(c, \mathbf{b})$. We can continue in this tradition via the device of normalizing consumption. Reciprocally, if a particular work of climate-policy scholarship has failed to normalize, and individuals are substantially heterogeneous with respect to relevant non-consumption attributes, we can see this as a shortcoming of the work that can be improved upon in subsequent research.

[5]Specifically, in footnote 4 immediately above, $v(.)$ would be a utility function representing the common preference structure, and $u(.) = u(c)$ applied to normalized consumption would be derived from $u(.)$ as per that footnote.

with either common or heterogeneous preferences, see generally Adler (2012, Chap. 3); Adler (2016a, b)].[6]

We can now write down the formula employed by the DU SWF. Let $W^{DU}(x)$ indicate the "social value" of outcome x: the numerical value assigned by the DU SWF to that outcome. $W^{DU}(x)$ is calculated as follows.

$$W^{DU}(x) = \sum_{t=1}^{T_{max}} \sum_{r=1}^{R} P_{tr} \times u(c_{tr}(x)) \times \frac{1}{(1+\rho)^{t-1}} \quad (5.1)$$

The parameter ρ, a positive number, is the rate of pure time preference, which defines the discount factor for time period t: $\frac{1}{(1+\rho)^{t-1}}$. Note that this discount factor decreases as time goes on. The discount factor for period 1, the present, is 1; the discount factor for period 2 is $\frac{1}{(1+\rho)}$; the discount factor for period 3 is $\frac{1}{(1+\rho)^2}$; and so forth.

It is typically assumed that well-being as a function of consumption takes the constant relative risk aversion (CRRA) form (Nordhaus 2007, 2008; Botzen and van den Bergh 2014; Stern 2007; Dasgupta 2008; Pindyck 2013). That is,

$$u(c) = \frac{c^{1-\eta}}{1-\eta} \quad (5.2)$$

with $\eta \geq 0$, $\eta \neq 1$ or $u(c) = \log c$ in the special case of $\eta = 1$. The parameter η of the CRRA well-being function captures the degree to which individuals (in their common preference structure) are risk averse with respect to consumption lotteries. A larger value of η means more risk aversion (Gollier 2001).

The slope of $u(.)$ at a given level of c is the "marginal utility" of consumption: the increase in well-being, per unit of consumption. With $\eta > 0$, marginal utility is decreasing: an increment of consumption at a greater level of consumption makes a smaller well-being difference than the same increment at a lower level of consumption. Moreover, as η increases, marginal utility decreases at a faster rate.

What exactly is the role of Eq. (5.1) in policy modelling? An "outcome" means a possible allocation of consumption and non-consumption attributes, or equivalently of normalized consumption, for each region in each time period. Ignoring uncertainty, we might assume that the status quo policy option of inaction ("business as usual") would yield outcome s, that some alternative policy would yield a particular outcome x, that a different policy would yield outcome y, and so forth. Policies can then be ranked in light of the outcomes to which they correspond, via Eq. (5.1).

---

[6]A different concern is that preferences, whether common or heterogeneous, may be poorly informed, irrational, "adaptive," or otherwise misshapen in ways that undercut their normative relevance. The response to this important worry is to analyze well-being in terms of preferences that are "idealized," e.g., fully informed and satisfying axioms of formal rationality. (Adler 2012, Chap. 3). To be sure, empirically estimating the preferences that individuals would have, if they met these idealizing conditions, is challenging.

More realistically, the status quo and various policy interventions are, each, probability distributions across outcomes. The DU SWF will require a more complicated formula. Whether policy a is better than b will depend both on the probability of each outcome conditional on a and b, and on the ranking of outcomes by Eq. (5.1). This further elaboration of the SWF approach is not discussed here [See Adler (2012, Chap. 7), Fleurbaey (2010), and Mongin and Pivato (2016) for general discussion of SWFs under uncertainty, and Adler and Treich (2015) for the case of climate policy. On climate change as a problem of risk assessment, see generally Gardoni et al. (2016)].

As already suggested in the brief remarks at the beginning of this chapter, the DU SWF is subject to two plausible (and logically distinct) criticisms. The first concerns the positive rate of time preference ρ. This creates an ethically arbitrary preference for individuals who exist earlier in time. Assume that two time-region pairs are identically situated in the status quo, save for temporal location. The population of region r at time t is equal to the population of region r^* at time t^*; and status quo per capita normalized consumption (and thus well-being levels) are the same too. We have the choice between one policy, which increases the well-being of each individual in region r at time t by an increment Δu; or a second policy, which increases the well-being of each individual in region r^* at time t^* by the same increment Δu. If time t is earlier than time t^*, the DU SWF prefers the first policy; if time t is later, it prefers the second policy; and the DU SWF is indifferent only if $t = t^*$.

But this seems quite unjustified. A more attractive SWF would be indifferent between the two policies regardless of whether $t > t^*$, $t < t^*$, or $t = t^*$. The individual well-being benefits are the same, as is the total benefit; and since starting point well-being levels are the same too, distributional considerations do not favor one policy over the other.

One defense of the time-preference factor is that *nondiscounted* utilitarianism may lead us to "sacrifice the present for the future" (Farber 2003; Nordhaus 2007, 2008; Weitzman 2007). Assume that there are positive returns to investment: a reduction in present consumption will produce a stream of future consumption benefits that is larger, in aggregate, than the reduction. If well-being is linear in consumption (i.e., the parameter η of risk aversion is zero), nondiscounted utilitarianism will always recommend the reduction—all the way to the point where current consumption is zero. However, the problem just described can be avoided without the resort to a time preference factor. Increasing the degree of risk aversion η mitigates the tendency to sacrifice the present for the future, as does the shift from utilitarianism to prioritarianism. (We will see some specific examples later).

A different defense of a positive rate of time preference is that policymakers are not, in fact, impartial between the generations (Nordhaus 2007, 2008; Weitzman 2007). Their actual attitudes are biased towards the present. For example, the U.S. Congress now is more concerned about present U.S. citizens than about U.S. citizens 200 years hence. This defense raises complicated issues regarding the relation between ethical decisionmaking, rational choice, and the SWF format, which will be briefly addressed below.[7]

The second and logically distinct criticism of the DU SWF is that it ignores the distribution of well-being. Assume that two regions, r and r^*, have the same population at some time t. In the status quo, per capita well-being levels are unequal. Individuals in region r have well-being level u, while individuals in region r^* are better off, at level u^*. Assume that one policy equalizes well-being levels with no loss, so that every individual in both regions has well-being level $u^+ = (u + u^*)/2$. A second policy equalizes well-being levels with some loss, so that every individual in both regions has a well-being level equaling $u^+ - \varepsilon$, with $\varepsilon > 0$. The DU SWF is indifferent between the status quo and the first policy, and prefers the status quo to the second policy.

But many will find this ranking problematic. A pure equalization of well-being, with no loss in total well-being, is—very plausibly—an ethical improvement. So the first policy should be preferred to the status quo. Moreover, the second policy should also be preferred to the status quo for ε sufficiently small.

The Nondiscounted Prioritarian SWF

The concept of "prioritarianism," much discussed in moral philosophy beginning with the work of Derek Parfit (Parfit 1991, 2012; Brown 2005; Holtug 2010, 2015; Porter 2012; Tungodden 2003; Williams 2012), is that well-being changes have greater ethical weight if affecting those at lower well-being levels. In other words, well-being has diminishing marginal moral significance: the ethical benefit of moving someone at well-being level u to level $u + \Delta u$ is greater than the benefit of moving someone at well-being level u^* to level $u^* + \Delta u$, if $u^* > u$. Those at lower well-being levels have ethical "priority"—thus the term "prioritarianism."

The NP SWF, which implements the concept of prioritarianism, and is abbreviated W^{NP}, is defined as per Eq. (5.3) [On the NP SWF, see Adler (2012, Chap. 5), Adler and Treich (2015), Adler et al. (2017)].

$$W^{NP}(x) = \sum_{t=1}^{T_{max}} \sum_{r=1}^{R} P_{tr} \times g(u(c_{tr}(x))), \quad (5.3)$$

with $g(.)$ a strictly increasing and concave function.

Note that this deviates from Eq. (5.1), for the DU SWF, in two respects. First, well-being numbers are inputted into a transformation function, $g(.)$, and then summed. Second, the time-preference parameter ρ is omitted.

---

[7]Two further defenses of the time-preference factor are that it reflects uncertainty about the future [as in its use by Stern (2007) to reflect extinction risk] and that it is required for well-defined sums of future well-being in the context of infinite time. For responses, see Adler and Treich (2015).

As illustrated in Fig. 5.1, the effect of the transformation function is indeed to give greater ethical weight to well-being changes affecting those at lower well-being levels. A closely related point is that the NP SWF is sensitive to the distribution of well-being. The total well-being in each outcome is not a "sufficient statistic" for ranking them.

Consider two individuals, one at a lower well-being level u_1, the second at a higher well-being level u_2. Because the g(.) function is strictly concave, a change in the first individual's well-being by amount Δu has a bigger impact on her g-transformed well-being than a change in the second individual's well-being by the same amount Δu. This also means that a pure transfer of well-being of Δu from the second individual to the first increases the value of the prioritarian SWF, i.e., the sum of g-transformed well-being numbers.

Return to this hypothetical case: Two regions, r and r^*, have the same population at some time t. In the status quo, individuals in region r have well-being level u, while individuals in region r^* are better off, at level u^*. Assume that one policy is such that all individuals in both regions have well-being level $u^+ = (u + u^*)/2$; while with a second policy, all individuals in both regions have well-being levels equaling $u^+ - \varepsilon$, with $\varepsilon > 0$. The NP SWF prefers the first policy to the status quo; it prefers the first policy to the second policy, but prefers the second policy to the status quo for ε sufficiently small.

Because the NP SWF does not incorporate a rate of time preference ρ, it is (appropriately) indifferent between policies where the affected regions are identical save for temporal position.

How shall we specify the transformation function $g(.)$? A strong axiomatic argument can be mounted that $g(.)$ should be a power function, as per Eq. (5.4) below.[8]

$$g(u(c)) = \frac{(u(c) - u(c^{zero}))^{1-\gamma}}{1-\gamma} \qquad (5.4)$$

with $\gamma > 0$, $\gamma \neq 1$ or $\log(u(c) - u(c^{zero}))$ in the special case of $\gamma = 1$.

Using Eq. (5.4), the transformation function is defined by two parameters: a priority parameter γ, and the parameter c^{zero}. Note that Eq. (5.4) can be further specified by defining $u(c)$ using the CRRA well-being function, as per Eq. (5.2) above. If so, the NP SWF is parameterized by three parameters: the priority parameter γ, c^{zero}, and the coefficient of individual risk aversion η. By contrast, the DU SWF with the CRRA well-being function is parameterized by two parameters: individual risk aversion η and the rate of time preference ρ.

The priority parameter γ captures the degree of concavity of the transformation function in Fig. 5.1. A larger value for γ means a greater degree of priority for worse off individuals. Parameter γ is a pure ethical parameter. Ethical intuitions about a normatively appropriate level of γ can be sharpened via various thought experiments, such as "leaky transfer" thought experiments. Assume that the ratio

---

[8]This axiomatic argument, relating to ratio invariance, and the meaning of c^{zero}, are elaborated at length in Adler (2012, Chaps. 3 and 5); Adler and Treich (2015); and Adler et al. (2017).

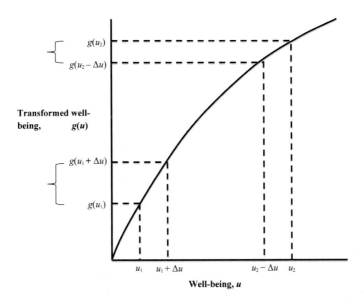

Fig. 5.1 A concave transformation function

between the well-being of two individuals is $K > 1$. Consider a policy that produces a small loss to the well-being of the better-off individual, with a benefit to the worse-off individual that is some fraction f of that loss. This "leaky transfer" is seen as a net improvement by the non-discounted prioritarian SWF if $f > (1/K)^\gamma$. Note that the acceptable degree of leakage $(1 - f)$ increases as γ increases.

Although the priority parameter γ looks mathematically similar to the risk aversion parameter η, the two are quite different. Risk aversion is a feature of individual preferences with respect to consumption lotteries. We identify an individual's risk aversion by looking at how she ranks consumption lotteries (with this ranking inferred either from her behavior or from her response to surveys). We are simplifying by assuming common preferences; but in reality preferences are heterogeneous and so too, then, is risk aversion. By contrast, the level of γ is specified by ethical reflection, concerning the appropriate degree of priority for the worse off.

The level c^{zero} is a consumption level so low that those in its vicinity come close to a kind of absolute priority over more fortunate persons. One natural thought is that c^{zero} is the *subsistence* level of consumption, below which ongoing existence is seriously at risk.

How Should We Choose Between the Two SWFs?

How should we conceptualize the choice of functional form of an SWF? What makes the NP SWF a "better" policy tool than the DU SWF, or vice versa? And who gets to make that choice?

The view adopted here, and well supported (although certainly not unanimously adopted) in the economics literature on SWFs,[9] is that an SWF is a tool for *ethical* decisionmaking. A decisionmaker adopts the *ethical* perspective (as opposed to a self-interested or parochial perspective) when she is impartial between all persons. It is the nature of ethical thinking to give each individual equal consideration just in virtue of his personhood.

The substantive content of ethical requirements is, of course, controversial. A decisionmaker needs to engage in *normative deliberation* in order to determine, as best she can, what she is ethically required to do. Normative deliberation involves a process of "reflective equilibrium," whereby the decisionmaker tries to synthesize her initial intuitions about general principles, and about what's ethically required in concrete cases, into a coherent ethical framework (Adler 2012, Chap. 1).[10] In defending the NP SWF (as opposed to the DU SWF or other frameworks), I am expressing the results of my own normative deliberation—in reflective equilibrium, this is the methodology for ethical thinking that I support—and articulating the considerations that lead me to this equilibrium point.[11]

A different view is that the specification of an SWF is "descriptive" [On the descriptive/normative debate within the climate literature, see Dasgupta (2008), Arrow et al. (2014)]. A given society "has" an SWF, and it is a descriptive question—a question of social fact—what that SWF is. But this view runs afoul of Hume's famous dictum, "no ought from is." An SWF is mean to provide policy *guidance*, to tell a decisionmaker what she ought to do. No combination of purely descriptive facts (including social facts) will suffice to yield a normative recommendation.

---

[9]The SWF was clearly understood by pioneering scholars, in particular Bergson (1948, 1954), Samuelson (1947, p. 221), and Harsanyi (1977, Chap. 4), as a tool for ethical deliberation. In the more recent literature, this view is at least implicit insofar as an axiom of "anonymity" is adopted (Weymark 2016; Adler 2012, Chaps. 1 and 5)—this axiom being a formal expression of the ethical norm of impartiality.

The anonymity axiom requires that the SWF be indifferent between a given list of well-being numbers for the population of concern, and all rearrangements of that pattern. Prioritarianism satisfies the anonymity axiom: greater weight is given to well-being changes affecting worse-off individuals—but because they are worse off, and not because of their names, identities, or other ethically irrelevant characteristics. Discounted utilitarianism satisfies the anonymity axiom with respect to rearrangements of well-being within each generation, but violates the axiom intertemporally. This is, indeed, why, the time-preference factor is ethically arbitrary. Adler and Treich (2015, p. 283, n. 7)

[10]Although the reflective-equilibrium methodology originates with John Rawls, who of course had specific views about justice and political liberalism (Rawls 1993, 1999), it can be understood as a general account of moral reasoning that is independent of those views.

[11]It should not be thought that the reflective-equilibrium methodology is solipsistic. A given decisionmaker's point of reflective equilibrium will surely be shaped by the intuitions and arguments of others. If I have at hand a particular decision to make, then, at the end of the day, *I* have to determine what *I* believe to be ethically required with respect to that decision—but I certainly can and, often, should engage in ethical deliberation and debate with others before arriving at this determination.

Moreover, in any modern, pluralistic society, ethical questions will be disputed by the citizens themselves. Some will be welfarists, believing that ethical requirements ultimately reduce to the allocation of well-being. Within the welfarist camp, some citizens will be inclined to utilitarianism, others to non-utilitarian versions of welfarism, such as prioritarianism. Yet other will reject welfarism, and may instead endorse views that include rights-based or deontological constraints on the maximization of good consequences understood in terms of welfare [On the range of ethical views, see Kagan (1998)]. Thus, in any modern pluralistic society, there will be no single SWF or even family of SWFs that is commonly endorsed by all (or a large majority of) citizens. The "descriptive" approach seeks after a social fact that doesn't exist.

It is certainly true that the *application* of an SWF depends in part on empirical facts. (For example, the use of an SWF to determine climate policy depends upon scientific facts regarding the environmental damage caused by carbon emissions). Moreover, an SWF might incorporate some parameters that are determined via empirical investigation. (For example, the η parameter, which calibrates the well-being function, is indeed descriptive rather than normative: it is an empirical question how risk-averse individuals are with respect to consumption lotteries). However, the initial choice of an SWF is an exercise in ethical deliberation. And it may, further, be the case that ethical deliberation is required to specify some of the parameters of a given SWF; these are ethical, not descriptive parameters (as is true, for example, for both the γ and c^{zero} parameters of the NP SWF).

Let's take as established the view of an SWF as a tool for ethical decision-making, one that is justified via a process of normative deliberation, that is endorsed by some deliberators in reflective equilibrium, and that will inevitably be controversial—since other deliberators will arrive at a different equilibrium point, endorsing a competing ethical approach. We can now ask a different question: What is the *legal* role of the SWF approach? More specifically: who in a given society is legally entitled to decide that the use of collective resources, and the issuance of laws and regulations governing members of the society, will be shaped by a given SWF?

In a democratic legal system, elected leaders are legally authorized to make controversial ethical decisions regarding the use of collective resources and the issuance of laws and regulations, subject to constitutional constraints and to being booted out of office (if those decisions are unpopular).[12] Thus arguments about the appropriate ethical framework for policy analysis, and specifically for climate policy, are directed to elected leaders and to the citizens who choose them.

It should now be conceded, indeed stressed, that ethical considerations are simply *one* input into decisionmaking by citizens and elected leaders. It is the rare citizen indeed who is solely motivated (in her day-to-day or voting behavior) by

---

[12]And, of course, this democratic legal setup has a strong ethical justification. There are strong ethical arguments, in light of human well-being, for creating and maintaining democratic legal systems that empower elected officials.

ethical considerations: by what she takes to be required from an impartial point of view, giving no more weight to her own interests than anyone else's. Self-interested considerations, or the welfare of her friends and family, will also play a decisional role, and likely a larger one. Nor is there anything irrational in this. The ethical perspective is rationally permissible, but so are others.[13]

Turning from the citizen to the elected official: although the official is rationally and legally permitted[14] to adopt a purely ethical perspective, she may rationally decide not to do so. Political feasibility is surely a rational consideration for elected leaders; the leader who is fully impartial between citizens and non-citizens is likely to find herself out of a job after the next election.

In short: the DU SWF and NP SWF, as conceptualized here, are candidate frameworks for determining what's ethically required, from a standpoint of impartiality; and what's ethically required, in turn, is only one ingredient in citizen and governmental decisionmaking.

We can now circle back to the question of time preference. Above, I criticized the DU SWF for incorporating a pure rate of time preference ρ, arguing that it is ethically arbitrary to give less weight to individuals who live later in time simply by virtue of their temporal position. This criticism (it seems to me) is pretty persuasive *if* the DU SWF is indeed conceptualized as a framework for ethical decisionmaking. But why not think of it differently: as providing guidance to an elected official from a *hybrid* perspective that takes account of *both* ethical considerations *and* parochial considerations (a concern solely for the interests of present-day citizens)?

I don't believe that this hybridized view of the DU SWF is workable. The constituency of the elected leader is current citizens. Thus a parochial perspective means ignoring the interests *both* of individuals in future generations (except insofar as current citizens care about those future individuals), *and* of current individuals who are non-citizens (except insofar as current citizens care about these current non-citizens). A hybrid framework would have both (steep) regional and temporal discount factors. Moreover, it is very hard to see why a hybrid framework would have a temporal discount factor with the exponential form of the DU SWF: whereby the discount factor for individuals in the next period is $\frac{1}{(1+\rho)}$; the discount factor for the period after that is $\frac{1}{(1+\rho)^2}$; and so on. Rather, the discount factor for an individual in a future period who is not a current citizen should reflect (a) the degree of concern of current citizens for this future individual, and (b) the extent to which

---

[13]On the rational permissibility of non-ethical behavior, see Sidgwick (1907); Scheffler (1982); Schelling (1995).

[14]Admittedly, there might be specific legal constraints prohibiting an elected official from adopting an ethical perspective. For example, a statute might specifically mandate that the President take a course of action that is inconsistent with ethical requirements as she sees them. Legislators are not bound by prior statutes, and thus will generally be freer to act ethically.

The recommendation to use some SWF in designing climate policy should, thus, be understood as an ethical recommendation addressed to elected officials—who will generally have some legal discretion to act upon such recommendations, with the precise scope of this discretion keyed to the official and to the legal context in which he operates.

this concern itself figures into the well-being of current citizens. Since current citizens are likely to care much more about their children and perhaps grandchildren than descendants further in the future, it is very hard to see how exponential discounting will be the upshot of (a) and (b).

The Ramsey Formula

Let's now compare the DU and NP SWF with respect to various specific topics of concern for climate economics, starting with the Ramsey formula (Adler and Treich 2015). The Ramsey formula is a key formula in the climate-policy literature and will be stated in a moment. It concerns the consumption discount rate, which I'll abbreviate as "λ". The reader should pay close attention, here, to different uses of the term "discount" and to the various distinct concepts that are tracked by these terminological variations. The consumption discount rate λ is *not* the same as the rate of time preference ρ. Rather, λ establishes an equivalence—in terms of impact on social welfare—between an increment of consumption at one time and an increment of consumption at a different time.

As typically formulated, the Ramsey formula ignores regional differences in per capita consumption. That is, it adopts the very simple premise that social welfare is a function of total consumption in each generation, assumed to be spread equally among all individuals alive regardless of regional location.

Let's first consider the *DU Ramsey formula*: the Ramsey formula as derived from the DU SWF [This is the version of the Ramsey formula that appears in extant climate scholarship—given the dominance of the DU SWF in this literature. See Stern (2007); Dasgupta (2008)]. Let c_1 and c_t denote per capita status quo consumption in, respectively, period 1 (the present) and period t.[15] Consider a small increment ΔC_t to total consumption[16] in period t. (Recall that consumption is normalized, so that ΔC_t might indicate a change in the flow of marketed goods and services, as caused by climate change, or a change in health or environmental quality expressed as a change in normalized consumption). We can now ask: what is the change ΔC_1 to total consumption in period 1 that has the very same effect on the value of the DU SWF as ΔC_t? What's the effect on social welfare (as calculated by the DU SWF) of a period-t consumption change—with this effect expressed not in social welfare units (which are abstract and hard to grasp) but instead in terms of an equivalent period 1 consumption change?

---

[15] The population in period 1 is P_1, and in period t is P_t. The formulas below are independent of these population sizes.

[16] Total consumption is the sum of individual consumption across the population. Thus the change to per capita consumption in period t is $\Delta C_t/P_t$.

The change in the DU SWF resulting from ΔC_t is approximately $(\Delta C_t)u'(c_t)/(1+\rho)^{t-1}$, with the prime denoting first derivative. The change in the DU SWF resulting from ΔC_1 is approximately $\Delta C_1 u'(c_1)$. Thus

$$\Delta C_1 \approx \frac{u'(c_t)}{u'(c_1)(1+\rho)^{t-1}} \Delta C_t \qquad (5.5)$$

We can now express the translation from ΔC_t to ΔC_1 in terms of a consumption discount rate. We can ask: what is the value λ^{DU} such that $\Delta C_1 = \Delta C_t/(1+\lambda^{DU})^{t-1}$? A bit of algebraic manipulation shows that:

$$\lambda^{DU} \approx \left(\frac{u'(c_1)(1+\rho)^{t-1}}{u'(c_t)}\right)^{1/(t-1)} - 1 \qquad (5.6)$$

Let's now assume that the well-being function $u(.)$ takes the CRRA form. Assume also that per capita consumption is growing at a positive rate d, i.e., $c_t = c_1(1+d)^{t-1}$. Then it can be shown that[17]:

$$\lambda^{DU} \approx \rho + d\eta \qquad (5.7)$$

This is the DU Ramsey formula. A change to period t consumption by ΔC_t has the very same effect on discounted-utilitarian social welfare as a change to period 1 consumption in amount $\Delta C_1 = \Delta C_t/(1+\lambda^{DU})^{t-1}$. The consumption discount rate λ^{DU} increases with the pure rate of time preference ρ, and with the product of the growth rate d and the coefficient of risk aversion η. Why this latter effect? As consumption grows, the marginal utility of consumption declines. A faster growth rate means a smaller amount of period 1 consumption change required to counterbalance a period t change. Moreover, a higher value of η means that marginal utility declines more rapidly for a given increase in consumption.

A parallel exercise can be carried out for the NP SWF, yielding the NP Ramsey formula. Consider a change to period t consumption of ΔC_t. What is the change to period 1 consumption ΔC_1 with the very same effect on social welfare—meaning now the value of the NP SWF—as ΔC_t? Let $g(.)$ be the prioritarian transformation function. The change in the NP SWF resulting from ΔC_t is approximately $\Delta C_t g'(u(c_t))u'(c_t)$, with the prime denoting first derivative. The change in the NP SWF resulting from ΔC_1 is approximately $\Delta C_1 g'(u(c_1))u'(c_1)$. We then have:

$$\Delta C_1 \approx \frac{g'(u(c_t))u'(c_t)}{g'(u(c_1))u'(c_1)} \Delta C_t \qquad (5.8)$$

---

[17] The derivation of this formula, and of the NP Ramsey formula below, uses the standard approximation that $\log(1+\Delta x) \approx \Delta x$ for Δx small.

As above, we now express the translation from ΔC_t to ΔC_1 in terms of a consumption discount rate λ^{NP}, such that $\Delta C_1 = \Delta C_t/(1+\lambda^{NP})^{t-1}$. If we assume a growth rate of consumption d, and CRRA well-being, we arrive at the NP Ramsey formula:

$$\lambda^{NP} \approx d\left(\eta + \gamma \frac{1-\eta}{1-(c_1/c^{zero})^{\eta-1}}\right) \qquad (5.9)$$

Various points are worth noting about the comparison of λ^{NP} and λ^{DU}. First, and most obviously, the two consumption discount rates are different. For a given change ΔC_t in period t consumption, the equivalent change to current consumption for purposes of discounted utilitarianism—the change ΔC_1 with an equal effect on social welfare as calculated with the DU SWF—may well be different from the equivalent change to current consumption for purposes of nondiscounted prioritarianism.

Second, both consumption discount rates are equal to $d\eta$ plus an additional positive term. In the case of λ^{DU}, that additional term is just the rate of time preference ρ. As time preference increases, so does λ^{DU}. With a larger weight on current well-being, a smaller change to current consumption will equilibrate (in terms of the DU SWF) a given change to future consumption.

In the case of λ^{NP}, the additional term is γL, with L positive.[18] Thus λ^{NP} increases as the priority parameter increases. This may be counterintuitive. Why should a greater degree of priority for the worse off mean a smaller concern for future as compared to present consumption changes? But recall that we are assuming positive consumption growth and no regional differentiation. So everyone in the present is worse off than everyone in future periods; and a larger degree of priority for the worse off translates directly into a greater concern for present consumption changes relative to future ones.

The relation between the degree of priority for the worse off, and the concern for future as compared to present consumption changes, become more complex with a regionally differentiated model (as we'll now see with our discussion of the social cost of carbon) and even more so with uncertainty about future consumption (a topic not pursued here).[19]

The Social Cost of Carbon

Imagine that a ton of CO_2 is emitted at present. This emission causes a stream of damages to future individuals, expressible as losses in their normalized consumption. Each time-region pair experiences a total consumption loss ΔC_{tr}. Each change

---

[18]As can be seen by rearranging terms in Eq. (5.9), $L = \frac{d(1-\eta)}{1-(c_1/c^{zero})^{\eta-1}}$

[19]As mentioned earlier, Adler and Treich (2015) discuss how uncertainty affects the application of the NP SWF to climate policy.

ΔC_{tr} produces a change to status quo social welfare—as calculated by the NP SWF or, alternatively, by the DU SWF.[20] The total impact on social welfare of the ton of CO_2 is the sum of these time-region specific changes.

We can now ask: what is the total reduction ΔC in current consumption with an equivalent social welfare impact as that of the ton of CO_2? This quantity is the social cost of carbon (SCC) [On the SCC, see generally van den Bergh and Botzen (2014), Greenstone et al. (2013), Pizer et al. (2014), Tol (2011)]. The magnitude of the SCC depends, of course, on the underlying SWF. SCC^{DU} and SCC^{NP} indicate the social cost of carbon as calculated, respectively, by the DU and NP SWFs.

But why calculate a social cost of carbon? As with the Ramsey formula (for the simpler case of a regionally undifferentiated model), it may be useful in terms of comprehension to express the abstract idea of a change to the value of the SWF in more concrete terms, as a change to current monetary expenditure.

A yet stronger rationale is this. The SCC gives guidance in determining an appropriate price to be imposed on the emission of a ton of carbon—be it a price imposed directly via a carbon tax or indirectly as the price in market equilibrium that emerges in a "cap and trade" system. Rational emitters will abate the emissions of an incremental ton of CO_2 if the dollar cost of abatement is less than the price, and will not do so if the dollar cost of abatement is more than the price. So the price should be set equal to the cutoff level of abatement cost (a cost to current consumption[21]) which is just equal to the benefit of abating a ton of CO_2. And the SCC is exactly this cutoff level.[22]

In a regionally differentiated model, the SCC (be it discounted-utilitarian or nondiscounted prioritarian) depends upon the regional incidence of the equilibrating change to current consumption. If one region has lower per capita consumption than a second, a given change to an individual's consumption in the first region will have a bigger well-being impact (given the declining marginal utility of consumption)[23] than the very same change to an individual's consumption in the second. This difference will be compounded by the NP SWF; a larger well-being impact for the poorer individual becomes a yet larger ethical impact, as a result of the priority for the worse off embodied in γ.

The regional incidence can be summarized in parameters $\pi_1, \ldots, \pi_R$, with π_r indicating the fraction of the reduction to current consumption that will be incurred by region r. These parameters can be thought of as measuring the incidence of a carbon price imposed by a given government or group of governments. For example, the incidence of a U.S. carbon tax might well be such that most of the lost

---

[20]In the case of the DU SWF, the change to social welfare is approximately $\Delta C_{tr} u'(c_{tr})/(1 + \rho)^{t-1}$. For the NP SWF, it is approximately $\Delta C_{tr} g'(u(c_{tr}))u'(c_{tr})$. These are analogous to the delta formulas used above in deriving the Ramsey formula, but with regional differentiation.

[21]To be sure, this is a simplification. Abatement costs might change investment as well as current consumption, and a more sophisticated calculation of the SCC would reflect that.

[22]The SCC can also be used to price carbon impacts for purposes of cost-benefit analysis (Greenstone et al. 2013). With appropriate distributional weights, cost-benefit analysis approximates an SWF (Adler 2016b).

[23]This assumes $\eta > 0$.

consumption flowing from the tax would be incurred by U.S. citizens. If so, the appropriate π_{US} to be used in calculating the level of a U.S. carbon tax would be close to 1, with π_r for other regions close to zero.

It should be reiterated that an SWF is best thought of as a tool for ethical deliberation. The SCC identified by a given SWF is, in turn, the reduction in current consumption ethically equivalent to the harms caused by a ton of carbon—ethically equivalent, as per that SWF. This SCC need not correspond to the price for carbon that an elected official would actually impose, given the balance of ethical and parochial considerations that motivate her. For example, a U.S. decisionmaker in calculating a carbon price might well ignore or downweight harms from carbon emission that are incurred outside the U.S. [On this issue, see Pizer et al. (2014)].

Note that SCC^{DU} is determined by the environmental impacts of CO_2 emissions (the stream of ΔC_{tr} damages for each time-region pair); by the incidence parameters $\pi_1, \ldots, \pi_R$; and by the parameters of the DU SWF, namely the risk aversion coefficient η and the rate of time preference ρ. SCC^{NP} is determined by the environmental impacts of CO_2 emissions; the incidence parameters; and the parameters of the NP SWF, namely η, the degree of priority for the worse off γ, and c^{zero}.

How does SCC^{DU} compare with SCC^{NP}? This is a question I have investigated with a team of collaborators (Adler et al. 2017). We have calculated SCC^{DU} and SCC^{NP} in the integrated assessment model RICE, and have also established analytical results regarding both SCCs. What follows are a few key results from this investigation.

With a low value of risk aversion η and zero time preference, SCC^{DU} is very large. This is an instance of "sacrificing the present for the future": extremely large reductions in current consumption are seen as justified to avoid the harms of a ton of CO_2. Adding a positive time preference reduces SCC from these extreme levels, but so does a larger value of η *or* shifting to the NP SWF. Even with a low value of η (and zero time preference), extreme values of SCC^{NP} are avoided by raising the priority parameter γ above zero.

The impact of ρ on SCC^{DU} is straightforward. Regardless of the incidence parameters (showing how the reduction in current consumption would be spread among the regions), SCC^{DU} decreases as time preference increases. By contrast, the impact of γ on SCC^{NP} is more complicated. If per capita consumption in every current region with non-zero incidence is less than per capita consumption in every future time-region pair affected by carbon damage, SCC^{NP} decreases as γ increases. (In this limiting case, those who bear consumption costs to mitigate future carbon damage are all worse off than all the beneficiaries—so increasing priority for the worse off means a lower SCC^{NP}). But if per capita consumption in some current region with non-zero incidence exceeds per capita consumption in some future affected time-region pair, SCC^{NP} may increase as γ does.

Third, in the central cases we analyzed,[24] SCC^{NP} is greater than SCC^{DU}. Nondiscounted prioritarianism tends to justify a higher carbon price than

---

[24]Here, we set c^{zero}, ρ, and γ to central values and compared the two SCCs as a function of η.

discounted utilitarianism. This proposition is not obvious. Removing the rate of time preference tends to raise the SCC, while inserting a priority parameter tends to lower it insofar as the current regions that would bear consumption costs are worse off than the future beneficiaries of carbon mitigation. How these effects balance out is hard to predict a priori. In the central cases analyzed, the net effect is to increase SCC.

Finally, and most generally, the level of the SCC is (in part) a *normative* question. First, the choice between the DU SWF and NP SWF (and their corresponding SCCs) is a normative question. Second, a decisionmaker who makes the threshold decision to adopt the NP SWF must engage in further normative deliberation to specify a social cost of carbon. SCC^{NP}, to be sure, has descriptive inputs (carbon impacts, incidence parameters, the coefficient of risk aversion), but also hinges on ethical parameters (γ and c^{zero}).

Optimal Mitigation

The Ramsey formula and the social cost of carbon focus on incremental changes. What is the impact on social welfare of a small reduction in future consumption, or of the incremental emissions of a ton of CO_2, with these impacts expressed in terms of an equivalent change to current consumption? A different question for climate analysts concerns *optimal mitigation*. Mitigation efforts at a given point in time use resources that could otherwise be channeled to consumption (at that time or later), but with the benefit that mitigation reduces harms to future consumption. What is the optimal path, over time, of consumption plus mitigation—the path that maximizes social welfare (as calculated by the DU or NP SWF)?

Optimal mitigation paths can be estimated using integrated assessment models such as RICE or DICE. But what drives the choice of path can be opaque. To illustrate more transparently how the choice between DU and NP measures of social welfare figures into optimal mitigation, I use a minimal two-period model (Adler and Treich 2015). The total size of the population in each period is P. In period 1, there is a stock of resources C that can be used either for consumption in period 1, or for investment (at rate d) for consumption in period 2. In this model, "investment" is a general idea that includes any way in which society can increase future consumption by refraining from current consumption. This includes the physical investment of resources that would otherwise be converted into current goods and services; the use of such resources to purchase carbon-reduction technologies; and the avoidance of future damage by leaving resources unused (keeping fossil fuels in the ground). All these pathways to reducing future damage, at current cost, are modelled as a tradeoff between consuming ΔC of the resources now, or $\Delta C(1 + d)$ later.

Total first period consumption is C_1, and total second period consumption is $C_2 = (C - C_1)(1 + d)$.

To model inequality, I assume that the population is divided into two groups. In the first period, consumption is split evenly among the groups. In the second period, one group receives fraction $\pi \leq \frac{1}{2}$ of total consumption, while a second group receives the remainder. To make the optimization tractable, I set $\eta = 0$ (utility linear in consumption) and $c^{zero} = 0$.

Let C_1^* be the optimal level of period 1 consumption, and $C_2^* = (C - C_1^*)(1 + d)$ be the optimal level of period 2 consumption. The DU SWF maximizes at a "corner solution." If the rate of time preference ρ is greater than d, the DU SWF consumes all the resources in the first period ($C_1^* = C$ and $C_2^* = 0$); if the rate of time preference ρ is less than d, the DU SWF consumes all the resources in the second period ($C_1^* = 0$ and $C_2^* = C(1 + d)$).[25] With no risk aversion, either the future is sacrificed for the present, or vice versa.

The NP result is quite different.

$$C_2^* = (\frac{C_1^*}{2})(1+d)^{1/\gamma}(\pi^{1-\gamma} + (1-\pi)^{1-\gamma})^{1/\gamma} \qquad (5.10)$$

This is a fascinating formula. There are subtle interactions between the growth rate d of invested resources, the priority parameter γ, and the degree of future inequality π. Note in particular that the ratio between C_2^* and C_1^* is *increasing* in π for $\gamma < 1$; invariant to π for $\gamma = 1$; and *decreasing* in π for $\gamma > 1$. For the prioritarian, decreases in future inequality may either increase or decrease current consumption depending on the degree of priority for the worse off (γ).

Conclusion

Prioritarianism is a well-developed concept in moral philosophy. And climate change is a central—perhaps *the* central—ethical issue of our time. Surprisingly, then, the prioritarian perspective has thus far played little role in the normative literature on climate change.

Prioritarianism is similar to utilitarianism in important respects. Both frameworks are consequentialist and welfarist. That is, both conceptualize the ethical status of choices (such as governmental policy choices) as determined by the ethical ranking of the outcomes that might result from these choices, plus the probabilities of the outcomes. The outcome ranking is, in turn, determined by the pattern of individual well-being (Adler 2012, Chap. 1).

However, prioritarianism departs from utilitarianism in giving greater weight to well-being changes affecting worse off individuals. Assume that Amy is better off than Bob in outcome x. Outcome y is identical to x, except that Amy's well-being in y increases by some amount (Δu) relative to x; and outcome y^* is identical to

---

[25]In the knife edge case where $d = \rho$, DU social welfare is indifferent to how much of C is invested.

x except that Bob's well-being increases by the very same amount. Then utilitarianism ranks y and y^* as equally good, while prioritarianism prefers y^*.

Relatedly, prioritarianism prefers to equalize a fixed total of well-being, while utilitarianism is indifferent to such equalization.

Utilitarianism and prioritarianism can be operationalized as tools for policy assessment using the methodology of the social welfare function (SWF). Indeed, an established research tradition in climate economics employs the discounted-utilitarian (DU) SWF to provide guidance on key questions. This chapter has introduced a different type of SWF—the non-discounted prioritarian (NP) SWF—and has compared the DU and NP SWFs with respect to three central topics in climate policy: the Ramsey formula, the social cost of carbon, and optimal mitigation.

These comparisons are meant to be illustrative, not definitive. The reader should be reminded that two important topics have not been discussed in this brief overview: variable population and uncertainty. Further, although substantial progress has been made on the theory of prioritarianism [see especially Adler and Treich (2015) on climate questions], much more empirical and modelling work is needed. What does giving priority to the worse off imply for climate policy? This is a question that, I hope, will excite the interest of climate researchers. Indeed, if prioritarianism is truly an ethical advance over utilitarianism (as I believe), then it is ethically important for the scholarly community to move beyond utilitarianism as the dominant framework for thinking about how we ought to mitigate and respond to climate change.

References

Adler MD (2009) Future generations: a prioritarian view. George Wash Law Rev 77:1478–1520
Adler MD (2012) Well-being and fair distribution: beyond cost-benefit analysis. Oxford University Press, New York
Adler MD (2016a) Extended preferences. In: Adler MD, Fleurbaey M (eds) The Oxford handbook of well-being and public policy. Oxford University Press, New York, pp 476–517
Adler MD (2016b) Benefit-cost analysis and distributional weights: an overview. Rev Environ Econ Policy 10:264–285
Adler MD, Treich N (2015) Prioritarianism and climate change. Environ Resour Econ 62:279–308
Adler MD, Anthoff D, Bosetti V, Garner G, Keller K, Treich N (2017) Priority for the worse-off and the social cost of carbon. Nat Clim Change 7:443–449
Arrow KJ et al (2014) Should governments use a declining discount rate in project analysis? Rev Environ Econ Policy 8:145–163
Bergson A (1948) Socialist economics. In: Ellis HS (ed) A survey of contemporary economics, vol 1. Richard D. Irwin, Homewood, pp 412–448
Bergson A (1954) On the concept of social welfare. Q J Econ 68:233–252
Blackorby C, Bossert W, Donaldson D (2005) Population issues in social choice theory, welfare economics, and ethics. Cambridge University Press, Cambridge
Boadway RW, Bruce N (1984) Welfare economics. Blackwell, Oxford
Botzen WJW, van den Bergh JCJM (2014) Specifications of social welfare in economic studies of climate policy: overview of criteria and related policy insights. Environ Resour Econ 58:1–33
Broome J (2008) The ethics of climate change. Sci Am 298:96–102

Brown C (2005) Priority or sufficiency ... or both? Econ Philos 21:199–220
Dasgupta P (2008) Discounting climate change. J Risk Uncertainty 37:141–169
Dasgupta P (2012) Time and the generations. In: Hahn RW, Ulph A (eds) Climate change and common sense: essays in honour of Tom Schelling. Oxford University Press, Oxford, pp 101–130
Farber DA (2003) From here to eternity: environmental law and future generations. Univ Illinois Law Rev 2003:289–336
Fleurbaey M (2010) Assessing risky social situations. J Polit Econ 118:649–680
Gardoni P, Murphy C, Rowell A (eds) (2016) Risk analysis of natural hazards: interdisciplinary challenges and integrated solutions. Springer, Cham
Gollier C (2001) The economics of risk and time. MIT Press, Cambridge, MA
Greenstone M, Kopits E, Wolverton A (2013) Developing a social cost of carbon for U.S. regulatory analysis. Rev Environ Econ Policy 7:23–46
Harsanyi JC (1977) Rational behavior and bargaining equilibrium in games and social situations. Cambridge University Press, Cambridge
Holtug N (2007) Animals: equality for animals. In: Ryberg J et al (eds) New waves in applied ethics. Palgrave Macmillan, Houndmills, pp 1–24
Holtug N (2010) Persons, interests and justice. Oxford University Press, Oxford
Holtug N (2015) Theories of value aggregation: utilitarianism, egalitarianism, prioritarianism. In: Hirose I, Olson J (eds) The Oxford handbook of value theory. Oxford University Press, Oxford, pp 267–284
Kagan S (1998) Normative ethics. Westview Press, Boulder
Kaplow L (2008) The theory of taxation and public economics. Princeton University Press, Princeton
Mongin P, Pivato M (2016) Social evaluation under risk and uncertainty. In: Adler MD, Fleurbaey M (eds) The Oxford handbook of well-being and public policy. Oxford University Press, New York, pp 711–745
Nordhaus WD (2007) A review of the *Stern Review* on the economics of climate change. J Econ Lit 45:686–702
Nordhaus WD (2008) A question of balance: weighing the options on global warming policies. Yale University Press, New Haven
Parfit D (1991) Equality or priority. Lindley Lecture, University of Kansas. Reprinted in Clayton M, Williams A (eds) The ideal of equality (2000). Palgrave, Houndmills, pp 81–125
Parfit D (2012) Another defence of the priority view. Utilitas 24:399–440
Pindyck RS (2013) Climate change policy: what do the models tell us? J Econ Lit 51:860–872
Pizer W et al (2014) Using and improving the social cost of carbon. Science 346:1189–1190
Porter T (2012) In defence of the priority view. Utilitas 24:197–206
Ramsey FP (1928) A mathematical theory of saving. Econ J 38:543–559
Rawls J (1993) Political liberalism. Columbia University Press, New York
Rawls J (1999) A theory of justice. Harvard University Press, Cambridge, MA
Samuelson PA (1947) Foundations of economic analysis. Harvard University Press, Cambridge, MA
Scheffler S (1982) The rejection of consequentialism. Clarendon Press, Oxford
Schelling TC (1995) Intergenerational discounting. Energy Policy 23:395–401
Sidgwick H (1907) The methods of ethics, 7th edn. Macmillan, London
Stern N (2007) The economics of climate change: the Stern review. Cambridge University Press, Cambridge
Tol RS (2011) The social cost of carbon. Annu Rev Resour Econ 3:419–443
Tungodden B (2003) The value of equality. Econ Philos 19:1–44
van den Bergh JCJM, Botzen WJW (2014) A lower bound to the social cost of CO_2 emissions. Nat Clim Change 4:253–258
Weitzman M (2007) A review of the *Stern Review* on the economics of climate change. J Econ Lit 45:703–724
Weymark J (2016) Social welfare functions. In: Adler MD, Fleurbaey M (eds) The Oxford handbook of well-being and public policy. Oxford University Press, New York, pp 126–159
Williams A (2012) The priority view bites the dust? Utilitas 24:315–331

Part III
Natural Hazards, Resilience and Mitigation

Chapter 6
Assessing Climate Change Impacts on Hurricane Hazards

David V. Rosowsky

Abstract This chapter summarizes recent work to examine whether there be any effect of postulated climate change scenarios on the hurricane (joint wind and rain) hazard. Considering a worst-case climate change scenario from the most recent IPCC report and region along the US coastline that saw the largest increase in sea surface temperature under that scenario, results show conclusively that there is an effect on the hurricane hazard. The results of event-based simulation can be used to statistically characterize the hurricane hazard (wind-only, or wind and rain). This information can inform decision-makers, planners, emergency managers, electric power or other utilities, transportation and other public works departments, insurers or other risk portfolio managers. Results from such analyses also can be used to evaluate the effectiveness of possible mitigation strategies to ameliorate expected impacts and moderate risks (or consequent losses) to an acceptable level.

Introduction

About five years ago my research group, which had been working on hurricane hazard characterization and risk analysis, asked the question, "could emerging climate change models be incorporated into an event-based hurricane hazard analysis to project changes in the hazard in the next, say, 100 years?" This seemed a reasonable question given the evolution and confidence emerging in climate change scenarios, for example in the IPCC reports (Pachauri 2008, 2014), and the reasonable expectation that there would be locations in which the long-term hurricane hazard would indeed be affected. In the context of civil infrastructure design, the 100-year projection periods in many of the scenarios is indeed long-term, on the order of twice the design life of most buildings in the US, for example.

D. V. Rosowsky (✉)
University of Vermont, 348 Waterman Building, 85 S. Prospect Street, Burlington 05405, VT, USA
e-mail: david.rosowsky@uvm.edu

Our event-based hazard modeling techniques had evolved in the years leading up to our inquiry, had been vetted through peer review by our respective scientific communities, and were being adapted by other researchers for their purposes. However they did not address climate change or contemplate model parameters that might be non-stationary as a result of climate impacts.

In reviewing both the IPCC reports and supporting documentation, and the underlying scientific papers describing both the climate change phenomena and the models created to capture and forecast the associated changes, we found that the most dominant variable in our event-based models (sea-surface temperature) was, in fact, one of the variables most affected by projected climate change in all of the scenarios. This therefore seemed a sensible place to start. We have since confirmed, through our own sensitivity studies and those by other researchers (e.g., Emmanuel 2008; Knutson and Tuleya 2004; Knutson et al. 2010), that capturing the changes in sea-surface temperature allows you to capture most of the impact of climate change (as reflected in current scenarios and models) on the hurricane hazard.

Our group has written extensively on the development and application of our models, all in the technical literature, over the last (nearly) two decades. These papers provide complete information on the development, validation, and application of the models. The interested reader is referred to these papers (Lee and Rosowsky 2007; Mudd 2014; Mudd et al. 2014a, b; Rosowsky et al. 2016).

In recent years, we have been invited to present our work to highly interdisciplinary groups, technical and non-technical, having interest in natural hazards, climate change impacts, and/or infrastructure resiliency. I presented our most recent work earlier this year at the *International Workshop on Climate Change and its Impact: Risk and Inequalities*, hosted by the University of Illinois. This chapter is based on that presentation and, as such, I have chosen to write it in a less technical style, hopefully making it more accessible to a broader audience of interested parties. The chapter starts with an overview of hurricane models, and then discusses how to incorporate climate impact into the hurricane hazard analysis using those models (considering both wind and rain). This is followed by a discussion of results (and their possible applications) from a simulation-based analysis to characterize the hurricane hazard considering the effects of projected climate change. Finally, some discussion of ongoing and future work is presented, followed by some final reflections.

Hurricane Risk Analysis (The Basics) and the Hurricane Model

The analysis of risk requires an understanding (model) of the hazard(s), and a representation (model) of the system's response when subjected to the hazard(s). The model output informs the assessment of the failure probability (or failure of the structure or system meet a specific performance requirement). Risk analyses may be coupled (in which the hazard and response are mathematically coupled in the analysis, and the failure probability is determined through a convolution integral) or uncoupled (in which the hazard and response are separated analytically).

Our work focuses on the hazard model, more specifically the characterization of the hurricane hazard for use in a fully-coupled hurricane risk analysis. Of interest is the effect of projected climate change scenarios (as presented by the IPCC, and focusing first on the effect of sea surface temperature rise) on the hurricane hazard model.

The hurricane model itself, which forms the foundation for the event-based simulation studies conducted by the author and his colleagues (as well as by other research teams) consists of: (a) a gradient-level wind field model, (b) a genesis and tracking model, and (c) a decay model.

The gradient-level wind field model is constructed using data from aircraft reconnaissance flights (at aircraft or gradient level) and information from the HURDAT database, the tropical cyclone (hurricane) database maintained by NOAA. Once defined, the wind speeds predicted through the gradient-level model can be modified to provide estimates of surface-level winds (at the level of the civil infrastructure of interest). It has been shown that well-formed (well-defined, intensive) hurricane gradient wind fields can be well represented as a vortex with translational movement (Fig. 6.1). The shape of the rotational vortex (surface air pressure) is a function of the distance from the hurricane eye, heading direction, air density, translational wind speed, Coriolis parameter, central pressure deficit, radius

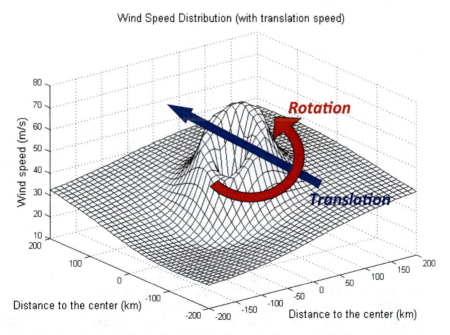

Fig. 6.1 Three-dimensional visualization of hurricane gradient wind field. *Source* Authors own

of maximum winds, and the Holland pressure profile parameter. These are defined elsewhere (Huang and Rosowsky 2000; Lee and Rosowsky 2007; Mudd et al. 2014a, b, 2016; Vickery et al. 2000).

The hurricane genesis and tracking model is based on information in the HURDAT database (Jarvinen et al. 1984), containing more than 150 years of records and information. This enables an occurrence rate to be established as well as parameters for a (Markov) model for storm tracking across the Atlantic basin. Finally, the overland decay model (how the storm intensity decays after it crosses over land) also is based on historical records of the rate of decrease in hurricane intensity (wind speeds) as the storm makes landfall and moves inland. As might be expected, the decay rate model is site-specific depending partially on the land features that serve to break up the storm. Storm decay is also the direct result of the hurricane being cut off from its energy source, the heat from the ocean. More information on the genesis, tracking, and decay models can be found elsewhere (Huang and Rosowsky 2000; Lee and Rosowsky 2007; Mudd et al. 2014a, b, 2016; Vickery et al. 2000).

Climate Change: Impacts on Hurricane Hazard

Our intention is not to argue any position on climate change or the rate at which climate change and associated impacts are occurring or may be expected to occur in the future. The scientific community has made the facts quite clear. The scientific data and projections are available for scrutiny or re-interpretation, validation or dispute. We stipulate to the scientific facts and the interpretations/projections presented by the global scientific community through the work of the IPCC (Pachauri 2014).

Warm air rising from the oceans is the major source of energy for hurricanes. In other words, heat is the 'fuel' for the hurricane 'engine'. These storms intensify, and become better organized (i.e., more symmetric) and therefore more efficient at capturing rising heat, as they pass over warmer waters.

The average global temperature has increased by 0.8 °C over the last 130 years, most (about 75%) of this change occurring in the last 35 years (since 1980). As a mesoscale meteorological event, it seems reasonable to contemplate that hurricanes may be affected by climate change. If the global temperature continues to rise, as is projected by the IPCC, it is certainly possible that hurricane hazard will be impacted. Our work seeks to examine the possible magnitude of that impact.

For our analysis, we selected a (high forcing function) scenario from the IPCC Fifth Assessment Report (Pachauri 2014). This scenarios (RCP 8.5) predicts 8.5 W/m^2 total radiative forcing by the year 2100, as compared to 1.6 W/m^2 in the previous report in 2005. The projected resulting increase in sea-surface temperature is shown in Fig. 6.2. We chose to focus our study on the northeast US coast. Figure 6.2 shows the water off the northeast US coast are projected to see some of the largest increases in sea-surface temperature along the eastern seaboard. This part of the eastern seaboard of the US

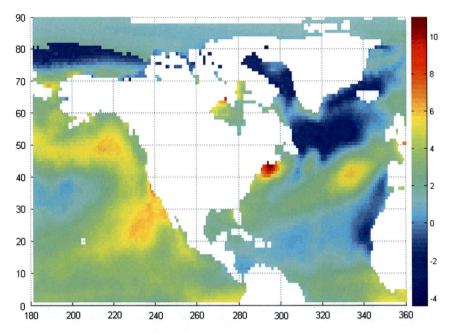

Fig. 6.2 Projected sea surface temperature change (degrees C.) under RCP 8.5, 2012–2100. *Source* Authors own

includes some of the largest cities (and populations) most vulnerable to intensified hurricanes in the future, as (1) the infrastructure is older, and (2) much of the infrastructure, new or old, was not designed for strong hurricane winds such as those that might be predicted to occur under climate change scenarios.

The Models: Wind and Rain

The wind field model (Huang and Rosowsky 2000; Huang et al. 2001; Lee and Rosowsky 2007) is based on models by Georgiou (1985) and Vickery et al. (2000); the strong track and central pressure models are based on the work by Darling (1991) and Vickery et al. (2000); and the decay model is based on the work by Vickery and Twisdale (1995). These models were adapted over the years to accommodate new information, additional features, site-specific and other conditions related to the study area under consideration (Lee and Rosowsky 2007; Mudd et al. 2014a, b, 2016; Wang and Rosowsky 2012).

Early in our work, Huang et al. (2001) showed that our models (informed by data from the HURDAT database) resulted in storm characteristics (and statistics of key parameters) that agreed well with actual records along the entire eastern seaboard (northeast, mid-Atlantic, southeast, and Gulf coasts). Specifically, comparisons were made between simulated and actual observed hurricane parameters (approach distance, arrival rate, translational speed, heading angle, and central pressure) along each section of the coastline. Actual data was available for milepost locations at 50 mile intervals along the entire coastline. This important work by Huang et al. (2001) provided the model validation needed to proceed in these early studies.

For the models to be used in regional loss estimation studies, which seemed a sensible next application, it would be important to properly characterize both the intensity *and* the size (or spatial extent) of the storm. Properly accounting for the size of the storm, the shape of the storm (relative intensity as you move further away from the eye), and both storm track and rate of decay all are necessary to properly estimate damage (and associated losses) over a spatial area. Mudd et al. (2014a, b, 2016) and Wang and Rosowsky (2012) expanded our group's earlier wind hazard characterization work to include both intensity (maximum wind speed, V_{max}) and spatial extent (radius of maximum winds, R_{max}). This resulted in bivariate (V_{max}, R_{max}) hurricane hazard characterization curves that demonstrated immediate potential for use in regional loss analyses, portfolio-wide vulnerability analyses, and even some performance-based engineering applications.

As hurricane hazard modeling and associated risk analyses (using the hazard models) matured, we sought to expand the bivariate (V_{max}, R_{max}) hurricane models to include rainfall, or more specifically rainfall rate (RR). This would allow the model to inform compound risk analyses considering both wind and rain. This could be of interest, for example, when looking at the risk of structures to wind and wind-induced rain intrusion, or when considering the possibility of local flooding when considering and planning for emergency response options.

The rainfall model is based on the R-CLIPER model by Tuleya et al. (2007). This rainfall intensity model is informed by data from NASA's Tropical Rainfall Measuring Mission Satellite Program (Huffman et al. 2010). Model validation and accuracy has been shown by Tuleya et al. (2007). Mudd (2014) later showed generally good agreement considering accumulated rainfall for a suite of 30 landfalling hurricanes (simulated vs. actual). The Tuleya model was then adapted by Mudd et al. (2016) to change from a deterministic to a stochastic (Weibull) rainfall intensity model. This allowed for the rainfall model to be fully coupled with the wind field model (through common parameters such as sea surface temperature and several of the gradient wind field parameters), enabling a fully coupled joint probabilistic (wind and rain) hurricane hazard model to be defined. In other words, the hurricane event model is incorporated into an event-based simulation framework (that considers the entire lifespan of the hurricane event, from genesis to overland decay), the model outputs of interest (e.g., V_{max}, R_{max}, RR) at landfall are recorded, and the joint wind-rain hurricane hazard is able to be statistically characterized. Our analysis is based on the simulation of 10,000 years of hurricane events making landfall along the US northeast coast (the study region).

Simulation: Hurricane Wind and Rain Events

Monte Carlo simulation was used to generate, propagate, and track hurricane (wind and rainfall) information, consistent with the models and information from the HURDAT database, described earlier. A complete description of the event-based hurricane simulation procedure, illustrated in Fig. 6.3, can be found elsewhere (Huang et al. 2001; Lee and Rosowsky 2007; Vickery et al. 2000). By running the complete simulation twice—once under current conditions and a second time with the increased sea-surface temperatures predicted under the IPCC scenario (RCP 8.5) —we were able to develop statistical information on landfalling hurricanes, specifically the joint wind-rain hazard, with and without consideration of possible climate impact. Selected results are presented in the following sections.

It is important to point out that the extreme wind climate along the northeastern US coastline is likely to be influenced *both* by tropical storms (hurricanes) and by extratropical storms (e.g., thunderstorms and other straight line wind events). As such, the extreme wind climate may not be characterized by hurricane wind speeds only. The results presented here can only be used to characterize the hurricane hazard (rather than the complete wind hazard) in this coastal region.

Results: Impact of Climate Change on the Hurricane Wind-Rain Hazard

The results of the event-based simulation can be used to statistically characterize the hurricane hazard (wind-only, or wind and rain). That is, event parameter sets (combinations of V_{max}, R_{max}, and/or RR) associated with different hazard levels (exceedance probabilities in given time periods) can be described. This information

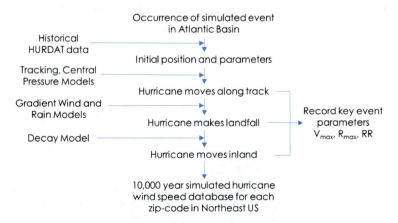

Fig. 6.3 Simulation procedure. *Source* Authors own

can inform decision-makers, planners, emergency managers, electric power or other utilities, transportation and other public works departments, insurers or other risk portfolio managers.

Results (event parameters at landfall) can be rank-ordered and fit with appropriate probability distributions, whether individually (marginal distribution) or as pairs (joint distribution). Pairwise (joint) probability densities are visualized as a surface in 3D, as seen in Fig. 6.4; it is not possible to visualize joint densities of more than two variables. Pairwise *hazard contours* (2D) can be extracted from 3D joint density functions, as seen in Fig. 6.5; and similarly *hazard surfaces* for three statistically dependent variables can be created and visualized (see Fig. 6.7). Hazard contours (or surfaces) are created at different hazard levels (e.g., x percent probability of exceedance in Y years) or their associated mean recurrence intervals. See Mudd et al. (2016), Wang and Rosowsky (2012) for more information and examples.

The results also can be used to compare hazard contours (or surfaces) with and without consideration of climate change. Figure 6.6 shows pairwise hazard contours for (V_{max}, R_{max}) and for (RR_{max}, R_{max}), in the year 2012 (present climate) and the year 2100 (under the RCP 8.5 climate change scenario). The results suggest that storms will intensify under the projected climate change scenario, becoming tighter (smaller R_{max}) and more intense (larger V_{max}), and the maximum rainfall rate (RR_{max}) will decrease. These findings are true at all but the lowest hazard levels (or shortest mean recurrence intervals). For example, in the case of the 10 year MRI, the results suggest storms become slightly larger. This is not significant for two reasons: (1) we have lower confidence (greater model uncertainty) in the smaller and less intense storms, and (2) we have greater interest in the extreme (high hazard or large MRI) events.

Figure 6.7 shows the hazard surfaces (three variable descriptors of hurricane events) with and without the effect of climate change. These can be rotated such that an in-plane projection is viewed, resulting in a set of two-parameter hazard contours, such as shown in Fig. 6.5. We have found the animation of the construction of these surfaces, and the ability to rotate the figure in three dimensions, have aided us tremendously in communicating both our statistical analysis and our findings.

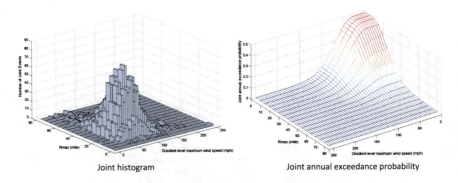

Fig. 6.4 Probabilistic description of bivariate hurricane event (V_{max}, R_{max}). *Source* Authors own

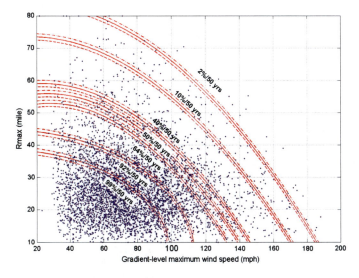

Fig. 6.5 Hazard contours (V_{max}, R_{max}). *Source* Authors own

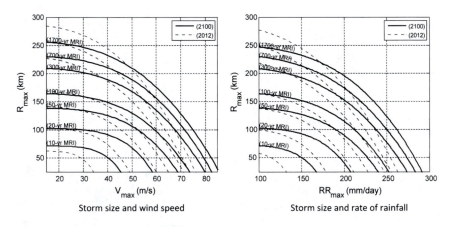

Fig. 6.6 Pairwise hazard curve examples, with and without impact of climate change. *Source* Authors own

How Can This Information Be Used?

We present two possible applications for the results of our work; others are discussed elsewhere (Mudd et al. 2016). First, we can use the results to examine the degree to which current design standards are adequate for conditions under the postulated climate change scenario. As an example, we consider the wind load provisions of the ASCE 7 standard for design loads on buildings (ASCE 2010).

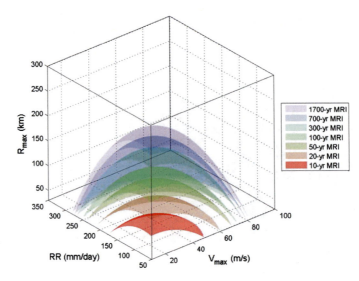

Fig. 6.7 Hazard surfaces (V_{max}, R_{max}, RR), future climate scenario (2100, RCP 8.5) shown. *Source* Authors own

Figure 6.8 shows the percent changes in the ASCE 7-10 design category II wind speeds under the RCP 8.5 scenario (at year 2100). Design category II corresponds to a 700-year mean recurrence interval. To create this figure, design wind speed maps (in this case, for a 700-year MRI) were constructed using simulation results at the zip-code level. These maps were then compared to the maps presented in ASCE 7-10 and percent differences in design wind speed were determined. What we see in this figure is that while many inland locations see little or no increase in design

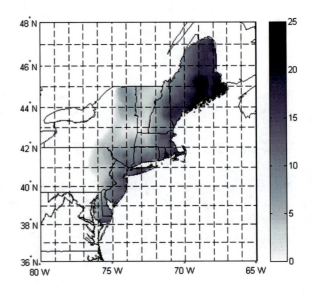

Fig. 6.8 Change in ASCE 7-10 design category II wind speeds. *Source* Authors own

wind speed, the majority of the northeast US coastline sees increases of 15% or more. This includes heavily populated locations having dense urban cores and vulnerable coastal infrastructure such as New York City and Boston. This part of the eastern seaboard of the US includes some of the most vulnerable infrastructure to intensified hurricanes in the future, as (1) the infrastructure is older, and (2) much of it was not designed for such strong hurricane winds. The implications for not properly accounting for anticipated such increases in design basis wind speeds (those in current load standards governing the design of buildings or infrastructure systems) in these regions are obvious, namely greater damage and economic losses at a minimum.

Second, we can use the results to create a suite of events (parameter sets) characteristic of a given hazard level. These can be used, for example, in a performance-based design or assessment application (designing for a specific performance requirements at specific hazard levels, or assessing performance of existing infrastructure similarly) or as input into a loss-estimation methodology (procedure to estimate cumulative losses, often over a geographic region, under a specific hazard scenario) (Sinh et al. 2016; Wang and Rosowsky 2012, 2013). Figure 6.9 shows the procedure used to construct Fig. 6.5 (e.g.) and further highlights a possible set of points (in this case spaced radially equally around the hazard contour) at each of two hazard levels. Each set represents a possible "suite" of events, characterized as a (V_{max}, R_{max}) pair, at the given hazard level. Therefore, looking at the outer ring of characteristic events (700-year MRI), one could use this suite of events, for example, as input to a regional loss-estimation model, or require

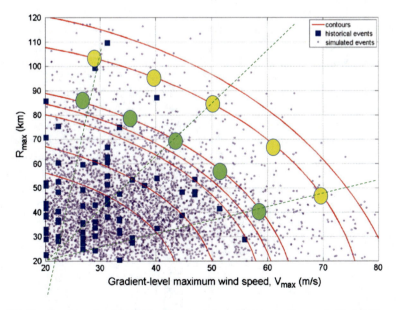

Fig. 6.9 Creating hazard-consistent hurricane event parameter sets. *Source* Authors own

that the design of an infrastructure component or system satisfy a certain performance requirement under each of these characteristic events.

Most importantly, we believe, is the potential for such analyses to inform code committees, planners, policy makers, and insurers about changes in the hurricane hazard and risk that can be expected under postulated climate change scenarios. Such work also can be used to evaluate the effectiveness of possible hard (e.g., structural) and soft (e.g., policy) mitigation strategies on ameliorating these expected impacts and moderating risks (or consequent losses) to an acceptable level.

A Look at Another Region Along the East Coast: The Mid-Atlantic

The work described previously was motivated by the fairly simple question, "would there be any effect of postulated climate change scenarios on the hurricane hazard?" To answer this question, we (a) chose a worst-case climate change scenario from the most recent IPCC report and (b) selected a study region along a section of US coastline that saw the largest increase in sea surface temperature under that scenario. The results show conclusively that there is an effect on the hurricane hazard.

We next sought to consider another US coastal region to see if similar effects were observed, despite less dramatic increases projected in sea surface temperature. We selected the North Carolina/South Carolina (NC/SC) mid-Atlantic coastline, as this represents a region of significant hurricane landfall activity and hence hurricane hazard. Unlike the northeast coastal region, however, there is less dramatic increase in sea surface temperature in the region immediately prior to landfall (see Fig. 6.2). There is, however notable increase in sea surface temperature across the Atlantic basin, over which storms generally travel as they approach the US. Our interest was whether or not this more moderate increase in temperature was sufficient, and existed over a sufficiently long fetch, to affect storm energy (intensity) and change the landfalling hurricane hazard in the NC/SC region.

While we have not yet completed the three-parameter analysis including rainfall rate, we are able to share results, with and without climate change, considering the wind speed (V_{max}) and storm size (R_{max}). The results, shown in Fig. 6.10, indicate that the storm intensity (maximum wind speed) increases due to projected climate change impacts, while the storm size (radius of maximum winds) increases slightly at the higher hazard levels (only). The tightening of landfalling storms (smaller R_{max}) that was generally seen along the northeast coast was not observed for those making landfall along the NC/SC coast.

6 Assessing Climate Change Impacts on Hurricane Hazards

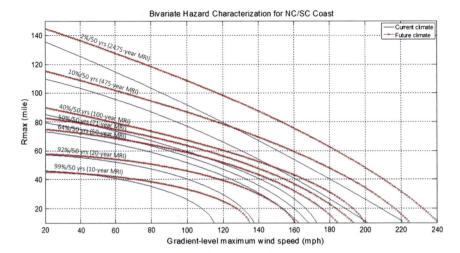

Fig. 6.10 Hazard contours (V_{max}, R_{max}) for the mid-Atlantic NC/SC coast. *Source* Authors own

Next Steps, Reflections, and Concluding Remarks

Our group set out to answer the relatively simple question: "Could emerging climate change models be incorporated into an event-based hurricane hazard analysis to project changes in the hazard in the next (e.g.) 100 years?" We had worked for several years to build event-based simulation frameworks (numerical procedures, and data-informed/calibrated models) without consideration of projected climate change impacts. Our question surrounded the feasibility of incorporating relevant information from emerging climate scenarios into our models to forecast changes in the hurricane hazard. We chose the RCP 8.5 scenario from the most recent IPCC report, as it represented a worst-case scenario. We chose to focus on a particular region (northeastern US coastline) as it coincided with the area of the Atlantic basin that was projected to have the largest increase in sea surface temperature.

We show that it is indeed possible for climate change scenarios to inform such event-based simulations models. In our work, the climate change parameter of interest is sea surface temperature, as heat from the ocean is the 'fuel' driving the hurricane 'engine'. While we have been able to show that climate change (characterized by increase in sea surface temperature) significantly impacts the landfalling hurricane hazard, our work has not been able to show that increase in sea surface temperature (alone) significantly impacts hurricane genesis or tracking across the Atlantic basin. There may well be other parameters (e.g., atmospheric vs. ocean) not accounted for in our models, however, that affect genesis and/or tracking as well as storm intensity.

It is quite reasonable to expect that researchers will continue this work, using more advanced models that capture more atmospheric and oceanic phenomena, and accessing more and better data. And of course, computational speeds will continue to increase making such analyses faster, easier, and more widely accessible. But even as we make incremental progress in these regards, it is unlikely to change the findings significantly. There is an impact of projected (and now broadly anticipated) climate change on the hurricane hazard in a period of time (100 years) relevant to decisions we must be making today. This is not a forecast that spans many centuries or many generations.

Recognizing, validating, affirming, communicating, debating, discrediting, or even just doubting climate change is not the point of our work. Indeed we stipulate, without comment, to a scenario from the most recent IPCC report. Most scientists agree that it will not be possible to reverse the trends brought about by human-induced climate change. Therefore, it must now be our goal to slow the rate and "best manage" (or adapt to) the changes that are taking place. The effective management of the risks created by the elevated hazard certainly will require robust and verifiable models on which to forecast changes to the hazards, and perhaps even the creation of new (or secondary) hazards as a result of these changes.

Our work can inform standards committees responsible for setting design-basis wind speeds in hurricane regions (for example), but it also can inform urban planners, building code officials, lenders, investors, insurers, utility companies, transportation officials, and the public. Such models can and should be used to inform infrastructure-related decisions, while the results can and should inform ongoing conversations around the expected impacts of climate change.

Acknowledgements The author is grateful to the following research group members for their many incremental and important contributions to our work in hurricane hazard modeling over the years: Zhigang Huang (Ph.D. student), Kyung Ho Lee (Ph.D. student), Lauren Mudd (Ph.D. student), and Yue Wang (Ph.D. student and post-doctoral researcher). Each built on the work of their predecessors, adding substantively to the state-of-the-art and the scientific literature. The author also is grateful to colleagues Chris Letchford, Frank Lombardo, Weichiang Pang, Peter Vickery, Peter Sparks, Ben Sill, Scott Schiff, and Tim Reinhold for their contributions and guidance over the years as this work evolved.

Writing about research in a non-technical style, or a less technical style than would more typically be employed to present scientific research, often requires an author to "stretch new muscles" to find their voice. The author acknowledges the work of Alan Alda and his team at the Alda Center for Communicating Science at Stony Brook University (and the affiliates network that includes the University of Vermont) for helping so many of us find that voice.

References

ASCE (2010) Minimum design loads for building and other structures, ASCE Standard 7-10. American Society of Civil Engineers, Reston, VA

Darling R (1991) Estimating probabilities of hurricane wind speeds using a large scale empirical model. J Clim 4(10):1035–1046

Emanuel KA (2008) The Hurricane-climate connection. Bull Am Meteorol Soc 89(5):10–20

Georgiou PN (1985) Design wind speeds in tropical cyclone-prone regions. Doctoral Dissertation, University of Western Ontario, London, Ontario

Huang Z, Rosowsky DV (2000) Analysis of Hurricane directionality effects using event-based simulation. Wind & Structures 3(3):177–191

Huang Z, Rosowsky DV, Sparks PR (2001) Hurricane simulation techniques for the evaluation of wind speeds and expected insurance losses. J Wind Eng Ind Aerodyn 89:605–617

Huffman GJ, Adler RF, Bolvin DT, Nelkin EJ (2010) The TRMM Multi-Satellite Precipitation Analysis (TMPA). In: Gebremichael M, Hossain F (eds) Satellite rainfall applications for surface hyrdology. Springer, Greenbelt, MD

Jarvinen B, Neumann C, Davis M (1984) A tropical cyclone data type for the North Atlantic Basin, (1886–1983): contents, limitations, and uses. Technical Report No. NWS NHC 22, National Oceanic and Atmospheric Administration (NOAA), Miami, FL

Knutson TR, Tuleya RE (2004) Impact of CO_2 induced warming on simulated hurricane intensity and precipitation: sensitivity to the choice of climate model and convective parameterization. J Clim 17(18):3477–3495

Knutson TR, McBride JL, Chan J, Emanuel K, Holland G, Landsea C, Sugi M et al (2010) Tropical cyclones and climate change. Nat Geosci 3(3):157–163

Lee KH, Rosowsky DV (2007) Synthetic Hurricane wind speed records: development of a database for hazard analysis and risk studies. Nat Hazards Rev 8(2):23–34

Mudd L (2014) A multi-hazard assessment of climatological impacts on Hurricanes affecting the Northeast US: wind and rain. Doctoral Dissertation, Department of Civil and Environmental Engineering, Rensselaer Polytechnic Institute, Troy, NY

Mudd L, Wang Y, Letchford C, Rosowsky D (2014a) Assessing climate change impact on the U. S. East Coast Hurricane hazard: temperature, frequency, and track. Nat Hazards Rev 15 (3):04014001

Mudd L, Wang Y, Letchford C, Rosowsky D (2014b) Hurricane wind hazard assessment for rapidly warming climate scenario. J Wind Eng Ind Aerodyn 133:242–249

Mudd L, Rosowsky D, Letchford C, Lombardo F (2016) A joint probabilistic wind-rainfall model for tropical cyclone hazard characterization. ASCE J Struct Eng, to appear

Pachauri RK (2008) Climate Change 2007—synthesis report, Contribution of Working Groups I, II, and III to the Fourth Assessment Report, Intergovernmental Panel on Climate Change (IPCC), Geneva, CH

Pachauri RK (2014) Climate Change 2014—synthesis report, Contribution of Working Groups I, II, and III to the Fifth Assessment Report, Intergovernmental Panel on Climate Change (IPCC), Geneva, CH

Rosowsky D, Mudd L, Letchford C (2016) Assessing climate change impact on the joint wind-rain Hurricane hazard for the Northeastern U.S. Coastline. In: Gardoni P, Murphy C, Rowell A (eds) Societal risk management of natural hazards, Springer, Heidelberg

Sinh HN, Lombardo FT, Letchford CW, Rosowsky DV (2016) Characterization of joint wind and ice hazard in Midwestern United States. ASCE Nat Hazards Rev 04016004

Tuleya RE, DeMaria M, Kuligowski RJ (2007) Evaluation of GFDL and simple statistical model rainfall forecasts for US landfalling tropical storms. Weather Forecast 22(1):56–70

Vickery PJ, Twisdale LA (1995) Wind-field and filling models for Hurricane wind speed predictions. ASCE J Struct Eng 121(11):1700–1709

Vickery PJ, Skerlj P, Twisdale L (2000) Simulation of Hurricane risk in the US using empirical track model. ASCE J Struct Eng 121(11):1700–1709

Wang Y, Rosowsky DV (2012) Joint distribution model for prediction of Hurricane wind speed and size. Struct Safety 35(1):40–51

Wang Y, Rosowsky DV (2013) Characterization of joint wind-snow hazard for performance-based design. Struct Safety 43:21–27

Chapter 7
Climate Change, Heavy Precipitation and Flood Risk in the Western United States

Eric P. Salathé Jr. and Guillaume Mauger

Abstract Current flood management, including flood control structures, land use regulations, and insurance markets, is adapted to historic flood risks, often using data from the past 100 years. In places where climate change will increase the flood risk outside the historic exposure, current management practices may not be adequate and losses could become increasingly catastrophic. For planning purposes, communities require scenarios of likely future flood inundation, which requires modeling the combined effects of sea level rise and changing peak flows along the relevant rivers, which in turn are derived from climate models and downscaling methods. In many regions, including the western United States, extreme precipitation is projected to increase with climate change, and these changes would have substantial impacts on flood risk. Simulating the effects of climate change on extreme precipitation presents substantial modeling challenges due to the complex weather dynamics of these events. Downscaling methods are critical to adequately incorporate the effects of climate change on extreme events and to simulate the response of local flood risk to these changes at the spatial and temporal scales most relevant to assessing community-scale risks from flooding. Statistical and dynamical downscaling is discussed and the implications of these methods for flood risk projections is evaluated. A case study is presented that illustrates three primary pathways for climate change impacts on a flood plain (sea level rise, reduced snowpack and higher intensity precipitation extremes) and illustrates the importance of methodological choices.

E. P. Salathé Jr. (✉)
School of Science Technology Engineering and Mathematics,
University of Washington Bothell, Bothell, USA
e-mail: salathe@uw.edu

G. Mauger
Climate Impacts Group, University of Washington Seattle, Seattle, USA
e-mail: gmauger@uw.edu

Introduction

Heavy rainfall during January 2009 cut off Interstate 5 and other major routes in Washington State, flooded major river drainages throughout the Pacific Northwest, and highlighted limitations in the ability of the Howard Hansen Dam in the Washington Cascades to protect $10–20 billion of assets and infrastructure and tens of thousands of people from flood risk (White et al. 2011). Rainstorms in France and Germany during May 2016 caused rivers to overflow their banks resulting in widespread flooding in Europe. Thousands of homes lost power, the River Seine overflowed in Paris threatening infrastructure and museums. The A.M. Best Company estimates insured losses from the floods at between EUR 0.9 billion and EUR 1.4 billion in France and around EUR 1.2 billion in Germany; especially in Germany, total economic losses were substantially higher. Given the probabilistic nature of rare events such as floods, it is not meaningful to ask whether climate change caused an event. Nevertheless, recent studies of extreme precipitation indicate a clear trend toward increased heavy precipitation in many regions worldwide (Easterling et al. 2000; Groisman et al. 2005) including much of the continental U.S. and parts of the West Coast (DeGaetano 2009; Mass et al. 2011). These trends are consistent with global projections of increased heavy precipitation with climate change (Tebaldi et al. 2006; Trenberth 2011). Positive trends are far from universal, however, and there is considerable uncertainty whether observed local changes are related to global climate change and will persist over the next few decades. In fact, natural variability is likely to continue to be a primary driver of local changes in the near term (Duliére et al. 2013).

Given the known link between human-induced climate change and heavy precipitation (Field et al. 2011; Hattermann et al. 2016) combined with uncertainty due to natural variability, recent events such as described above raise the critical question of whether floods are becoming more frequent as a result of climate change and whether society can expect flood risks to increase over the next few decades. Current flood management, including flood control structures, land use regulations, and insurance markets, is adapted to historic flood risks, often using data from the past 100 years. In places where climate change will increase the risk outside the historic exposure, current management practices may not be adequate and losses could become increasingly catastrophic.

Heavy precipitation in winter storms and snowmelt in spring are the primary causes of high flows and flood risk in much of the western United States (Neiman et al. 2008a). Flood risk is affected by a combination of factors, and modeling these processes requires high spatial resolution to represent the effects of mesoscale weather systems and terrain on extreme events and snowpack dynamics. A severe flooding event in Washington and Oregon in December 2007 illustrated the complex weather and climate factors typical of flood events in the western United States. A sequence of three storms hit the area over three days, the first two produced saturated soils and substantial snow cover in both the lowlands and mountains. The third storm was a so-called atmospheric river event popularly known as a "Pineapple Express", a common winter storm system that directs warm, moist tropical air to the US West Coast resulting in substantial precipitation and snowmelt. This combination of antecedent conditions and a warm tropical weather event resulted in considerable flooding and mudslides across the region with six counties declared federal disaster areas. Thus, to adequately project future changes in the flood risk due to extreme weather events like those described above, sophisticated climate and hydrologic models must be used that can represent the influences of global climate change, extreme and localized weather systems, antecedent conditions of snow and soil moisture, the flow of water across the land surface and in rivers, and the hydraulic effects of sea-level rise. To this end, we have developed methods, described below, that are applicable at the geographic and temporal scales necessary to inform community planning and response to changes in flood risk. Here, we review several of the important climatic effects that must be considered, outline the methods for simulating flood risk, and provide some case examples where these have been applied. Our focus is on the western United States, but the approach could be generalized and applied in many places worldwide.

Climatic Drivers of Flood Risk

Snowpack Effects

The timing of streamflow in mountainous rivers and streams is strongly controlled by the temperature and the seasonality of precipitation across the watershed that supplies the river. Basin temperature affects the geographic and elevational distribution of snow, and the proportion of winter precipitation falling as snow or rain within a river basin has a strong effect on streamflow. Especially in the Western US (Elsner et al. 2010), this temperature control on snowpack has become a critical means to categorize different rivers according to the seasonal timing of streamflow. Generally speaking, warmer river basins are found at lower elevations and along the coasts while cold basins drain the highest peaks of the Cascade, Sierra, and Rocky mountain ranges. For *rain dominant river basins*, the basin is generally too warm to accumulate snowpack and streamflow is coincident with precipitation. Typically, this results in a single maximum of streamflow and flood risk coincident with the heaviest rainfall in fall. In colder *snow dominant basins*, winter precipitation accumulates as snowpack, and streamflow peaks in spring and summer when the snowpack melts. In these basins, heavy fall precipitation events are essentially absorbed into the snowpack, reducing flood risks. Floods may occur, however, when snowpack melts rapidly into the river during spring. Intermediate temperatures result in transitional *mixed rain-snow basins,* which experience two peaks, corresponding to the fall rain-driven flow and the spring melt. Apart from any changes in precipitation, the effects of temperature can have profound impacts on flow characteristics and the risk of high flows by reducing the flow from melting snow and increasing the rain-driven flow (Hamlet et al. 2013). In addition to the shift towards higher flows in winter, rain-driven flow is more subject to intense precipitation events, resulting in an additional increase in flood risk (Tohver et al. 2014).

Heavy Precipitation

In the words of the Intergovernmental Program on Climate Change (IPCC), evidence for a warming trend in global temperatures is now "unequivocal" (IPCC 2007). The observational evidence for changes in the frequency, duration, and intensity of extreme precipitation events, however, suggest that increases are "likely", indicating relatively lower confidence that changes have already occurred. At a regional scale, observational and modeling evidence suggest that a trend toward increased precipitation in the western US is only recently emerging (Duliére et al. 2013). The reduced confidence in extreme precipitation trends relative to temperature trends stems from the small number of extreme events in the historical record and the influence of natural variability on short-term trends in extreme events (see for example, Warner et al. 2012). Therefore, it is difficult to attribute recent trends at a single location to an anthropogenic influence on the climate. However, research suggests that over large regions, which aggregate many individual weather events and improve statistical sampling, the observed trend to more frequent extreme events can be statistically attributed to the warming of the climate system with increased greenhouse gas emissions (Min et al. 2011). For example, a study including the entire region of the United Kingdom found that local trends may be detectable at that geographical scale in the next 20 years (Fowler and Wilby 2010). Furthermore, there are strong theoretical reasons to expect increases in heavy precipitation in a warming climate, since warmer air will be able to transport more water vapor into storm systems, following the Claussius-Clapeyron scaling of the saturated vapor pressure with temperature (Pall et al. 2007). Thus, warming would tend to increase the moisture available for precipitation in extreme events (Trenberth 2011) and the intensity of both wet and dry extremes (Held and Soden 2006).

As with the December 2007 event described above, heavy precipitation along the west coast of North America depends on well-known weather patterns (Colle and Mass 1996; Garvert et al. 2007) associated with atmospheric river events (Neiman et al. 2011; Warner et al. 2012). Atmospheric rivers are storm systems that produce an intense stream of warm moist air flowing to the east from the subtropics to the mid-latitudes. They form over the ocean all around the globe in both north and southern hemispheres and produce heavy precipitation along the west coasts of major continents. Atmospheric rivers are controlled by the large-scale circulation patterns in the atmosphere, occurring along waves in the mid-latitude jet stream. Projected changes in the jet stream (Chang 2007; Salathé 2006; Ulbrich et al. 2008) or atmospheric rivers themselves (Leung and Qian 2009; Neiman et al. 2008a, b) would therefore have substantial implications for future extreme precipitation events. In particular, climate models project an increased risk for more frequent extreme precipitation in the western US by the second half of the 21st century (Avise et al. 2009; Fowler et al. 2007) with more intense atmospheric rivers along the west coast (Dettinger 2011) and local intensification in areas of complex terrain (Salathé et al. 2010).

A number of observational and modeling studies find that, at both regional and global scales, total precipitation responds less to anthropogenic climate change than does heavy precipitation (Allen and Ingram 2002; Fowler et al. 2010; Frei and Schär 2001). For example, regional climate model simulations of total precipitation in current and future climate of the western US (Duliére et al. 2013) show inconsistent responses to climate change among three different climate models and across the western U.S. (Fig. 1, top row). However, for precipitation on days exceeding the historic 95th percentile, a more robust increase is found (Fig. 1, bottom row), and all three models show an increase over most of the western U.S. These results suggest that (1) processes strongly linked to climate change (e.g. thermodynamic processes such as moisture convergence, adiabatic lapse rate, large-scale circulation patterns) result in heavier precipitation during rainy events, (2) these effects are robustly simulated across the climate models, and (3) natural climate variability (e.g. El Niño, PDO) has a greater effect on the total annual precipitation. This natural variability will likely continue as in the past, dominating any trends in

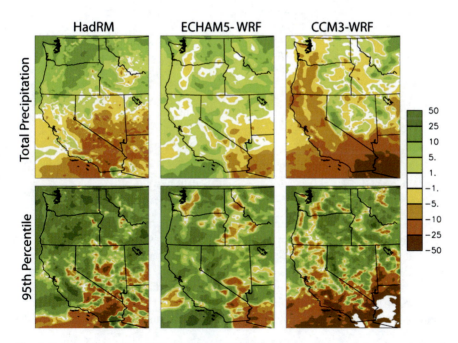

Fig. 1 Percent change from 1970–1999 to 2030–2059 in precipitation on all days (top row) and for days exceeding the historic daily 95th percentile (bottom row). Each column shows results for one of three regional climate model simulations. *Source* authors own

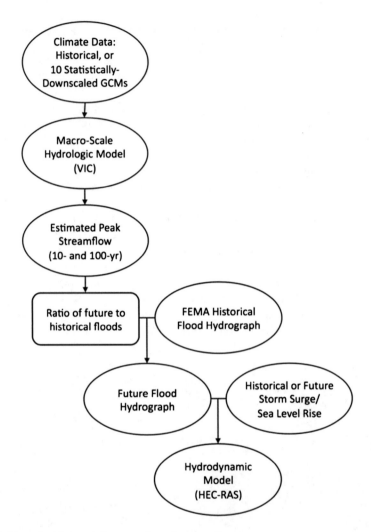

Fig. 2 Flow chart for modeling procedure. *Source* authors own

annual total precipitation (Mote and Salathé 2010). Similarly, Warner et al. (2015) found increases in winter mean precipitation off the west coast of the United States at a rate of about 5% per degree Celsius and changes of 10% or more per degree for atmospheric river events. The natural year-to-year variability in precipitation would overwhelm the modest increase in winter mean precipitation, and this change would not be discernible over the next century. The larger sensitivity of rainfall in atmospheric rivers, however, would emerge from natural variability in a few decades (Fig. 2).

The processes described above consider heavy precipitation at a very large, continental scale. However, the flooding from heavy precipitation impacts very small regions, often along only a few rivers and a few cities or counties. Mountainous terrain can produce localized intensification of precipitation, for example as air is lifted over windward slopes or converges in the lee of mountains with important implications for municipal agencies managing the flooding impacts of heavy precipitation. Alterations in heavy precipitation events associated with climate change include (1) the temporal and spatial extent of heavy precipitation, (2) the temperature anomaly and freezing level associated with storms, (3) the likelihood of storms occurring with specific antecedent conditions (e.g. high soil moisture or snow cover), and (4) the orientation of storms relative to local terrain. From the local perspective, these effects could have profound implications for the flood risks from extreme events that go beyond simple changes in the amount of rainfall, requiring an integrated approach to precipitation and surface hydrology.

Since shifts in the intensity and frequency of heavy precipitation or changes in storm characteristics (e.g. physical extent, duration, tracking relative to terrain) can have a substantial effect on streamflow and flood risk, changes in extreme events would have greater impacts in many regions than changes in total precipitation. The regional climate scenarios discussed here have been specifically designed to represent the response of the local flood risk to these changes at the spatial and temporal scales most relevant to assessing community-scale risks from flooding and to guide decision making in the face of these risks.

Methodological Approaches

To develop estimates of future flood risk from the information provided from global climate models, a number of additional modeling steps must be taken. The objective is to develop scenarios of future risk of flooding using the same parameters and nomenclature used by communities in planning for flood events today. Global climate models (GCMs) are the primary tool for simulating the future climate. Due to the computational constraints of simulating the global coupled atmosphere-ocean-land system, however, global models typically use a grid spacing of 100–200 km or more. While adequate to represent the climate system dynamics responsible for global climate change, information not produced at the spatial scale of the impacts of climate change, typically under 10 km. Thus, the information provided by the global models must be regionalized or downscaled to project the implications of changes in the large-scale environment on local conditions. We have conducted studies of flood risk and climate change using two methods to downscale the information from Global Climate Models, (1) statistical downscaling methods (Tohver et al. 2014) (see http://www.hydro.washington.edu/2860/) and a (2) dynamical downscaling using the Weather Research and Forecasting regional climate model (WRF) (Duliére et al. 2013) Both methods produce consistent high-resolution regional climate data, specifically for precipitation and temperature, which are used to model snowpack and river flows. Data from both products are used in parallel for projecting future flood risks in order to maximize the advantages of the two approaches, as discussed below. Our research has shown that regional climate models project very different future flooding scenarios than statistical downscaling, (Salathé et al. 2010, 2014). The fundamental reason for this difference is that statistical downscaling methods are designed to preserve the rate of change of temperature and precipitation simulated by the global model. Dynamical downscaling, by contrast, represents weather processes that are not resolved by the global models, which could create strong local differences with the global model result. Given the uncertainties in projecting the future climate and limitations of our current state of knowledge, it is not clear which of these approaches provides the best practice in a given application. However, by using both approaches and understanding where and why they diverge, one can gain important insight into the weather and hydrologic processes that contribute to future flood risks.

To generate quantitative flood risk scenarios, the downscaled daily temperature and precipitation projections are used as input to a distributed hydrologic model to simulate daily surface runoff and resulting streamflow volumes. The resulting regional climate and hydrologic scenarios have since been adopted by researchers at the USACE, USFS, and other regional agencies for assessing future flood risks and vulnerabilities (see, for example, Hamlet et al. 2013; Rybczyk et al. 2016).

Statistical Downscaling

The statistical downscaling method used in Tohver et al. (2014) is based on a gridded historical time series of temperature and precipitation from 1916 to 2006. In this dataset, station observations were interpolated onto a 0.0625-degree (1/16-degree) grid with elevational corrections applied to account for the effects of topography (Elsner et al. 2010; Hamlet et al. 2013). Since there are relatively few station observations in areas of high terrain, simple interpolation between stations would produce inaccurate results if the stations are at a lower elevation than the area between them. Daly et al. (1994) developed a sophisticated empirical model of the effects of topography—slope, aspect, and elevation—on the variation in temperature and precipitation with elevation, which is applied in the interpolation process to better represent temperature and precipitation in mountainous regions. To represent future climate conditions based on a global climate model simulation, the historic observations at each grid cell are perturbed to reflect the shift in the monthly-mean probability distribution projected by the global model for that location. Thus, the downscaled climate change scenarios repeat the historic sequence of daily weather, but with the projected future monthly statistics. In this approach, both the mean and higher moments of the observed data change in response to monthly GCM projections, and these changes vary spatially. Although the downscaling represents changes projected by the GCM, the future daily time series behavior and cross correlations of precipitation and temperature, and also the size, location, and inter-arrival time of storms match those in the historical record. As a result, a winter storm event or summer dry spell in the future will have the same location, spatial extent, and duration as its occurrence in the historical record, but the intensity of individual events are scaled to match the change in the monthly values projected by the GCM simulations. These limitations reflect the current lack of confidence in the ability of most global climate models to adequately simulate current and future daily weather statistics such as the time between storms, and until recently, most global modeling centers provided only monthly-mean climate data. Therefore, to aggregate the results from as many global models as possible in an even-handed way, the downscaling is based upon the change in monthly-mean climate, with the historic daily variability preserved. A smaller set of global models do adequately simulate daily weather statistics, and these may be selected for more detailed downscaling using a regional climate model as described below.

By design, the statistical method does not represent all the mechanisms that could affect hydrologic processes in the future. In particular, the characteristics of intense weather events could change in ways the global model does not simulate due to regional-scale atmospheric processes such as deep convection triggered by the terrain. For example, for a north-south oriented mountain range exposed to mid-latitude westerly winds, such as the Cascade Range of Washington and Oregon, precipitation may increase more on the western side than on the eastern side due to the higher intensity of rainfall on the windward slopes and the suppression of precipitation by the rain shadow. Statistical downscaling, therefore, produces regional scenarios of climate change that incorporate primarily the continental-scale and seasonal changes in temperature and precipitation, but not the effects of smaller scale weather processes and terrain interactions. With regard to changing flood risk, for the western U.S., the consensus of global models indicates a shift to more autumn precipitation and warmer autumn temperatures during the next century. The downscaling method allows a detailed simulation of the regional impacts of these shifts in the climate with the greatest confidence.

Regional Climate Model

The dynamical downscaling used for the case studies we discuss here were performed for a northwestern U.S. geographical domain using the Weather and Research Forecasting (WRF) community mesoscale model (Salathé et al. 2008, 2010; Zhang et al. 2009). The domain covers the northwestern U.S. encompassing Washington, Oregon, Idaho, northern California, and southern British Columbia with a spatial resolution of 12 km. Global climate model results obtained from several research centers are used to provide boundary conditions for the regional simulation (please refer to Salathé et al. (2010) for technical details of the regional modeling methods). This kind of simulation, where a coarse-resolution global model is used as input to a high-resolution regional model, is frequently referred to as dynamically downscaling the global model. The dynamical downscaling is performed for several global climate model simulations from the Climate Model

Intercomparison Projects (CMIP3 and CMIP5) (Meehl et al. 2007; Taylor et al. 2011). The Climate Model Intercomparison Project is an organized effort of the international climate modeling community under the World Climate Research Programme (WCRP) beginning in 1995. The project established a standard experimental protocol for studying the output global climate models and to support the Intergovernmental Panel on Climate Change assessments. Nearly all climate modeling centers around the world participate and perform climate simulations of the historic climate and projections of the future climate using common standards and data processing. Future climate projections are performed for a number of scenarios of future climate emissions reflecting a range of greenhouse gas emission pathways. CMIP3 and CMIP5 produced a large ensemble of climate model simulations that could be quantitatively compared against each other and against observations in order to select the most appropriate models for climate impacts studies (Mote and Salathé 2010; Rupp et al. 2013).

Although the regional climate model represents important terrain features and the mesoscale structure of storms that control flooding in rivers better than global models, deficiencies in both the global forcing fields and the regional model remain and introduce biases in the simulated climate variables (Christensen et al. 2008; Wood et al. 2004). Furthermore, to obtain acceptable hydrologic simulations, the simulated temperature and precipitation require additional downscaling from the 12 km WRF grid to the 0.0625-degree (approximately 6-km) grid used for hydrologic modeling as described in detail in Salathé et al. (2014). Briefly, the bias correction is based on a mapping between simulated and observed probability distribution functions to remove bias, while largely preserving local climate signals and time series behavior in the simulation. These steps remove systematic bias in simulated meteorological variables resulting from combined bias in the large scale forcing and WRF simulations.

Hydrologic Model

In order to simulate surface runoff and associated flow in the streams and rivers that collect runoff, the projected climate scenarios are used as input to a surface hydrologic model. In our work, we have used the Variable Infiltration Capacity (VIC) model (Liang et al. 1994) implemented 0.0625-degrees (Elsner et al. 2010). VIC is a macro-scale, fully-distributed hydrologic model that solves the water and energy balance at each model grid cell, producing (among other water balance variables) daily time step runoff and baseflow at each model grid cell. A separate streamflow routing model (Lohmann et al. 1996), is used to generate daily time step streamflows at various river locations from the simulated runoff and baseflow. The VIC model requires daily inputs of total precipitation, maximum and minimum air temperature, and mean wind speed.

The issues discussed above regarding rain and snow effects on river flow are important in understanding the uncertainties in hydrologic simulations. For rain-dominant basins, uncertainties in simulated hydrologic extremes are mostly related to uncertainties in simulated precipitation, which is obtained from the statistical downscaling or regional climate model. in regions where snow is an important part of the hydrologic cycle, however, uncertainties are related to the simulated temperature and to the snow pack simulation from the hydrologic model.

Case Study: Inundation Scenarios and Decision Support

Over the past decade, numerous studies have estimated changes in river flooding, sea level rise, and storm surge; however, few studies have quantified their combined impacts on flood risk. Here we present a case study that assessed the impacts of sea-level rise, storm surge and changing peak streamflows on flood inundation in the lower Snohomish River (for more information, refer to the project report, https://cig.uw.edu/wpcontent/uploads/sites/2/2014/11/FinalReport_CIG_TNC_Snohomish-20141209.compressed.pdf). Flood risk in the lower Snohomish River, located in the Puget Sound Region of Washington State, is impacted from the downstream marine boundary by storm surge and sea level rise (SLR), and from the upstream freshwater boundary by seasonal changes in river flow and hydrologic extremes. Building on previous work on the nearby Skagit River basin (Hamman et al. 2016), we developed projections of changing inundation in the lower Snohomish based on changes in these two boundary forcings. Results from this work have been incorporated into a decision support tool developed by The Nature Conservancy (TNC), designed to support multi-objective floodplain management by partners across Puget Sound. The goal of multi-objective floodplain management is to coordinate flood loss reduction measures with other community needs and goals for the floodplain; it depends on communication between different parties and on the technical and financial help from government agencies and private organizations.

For planning purposes, communities require scenarios of likely future flood inundation—that is, the areas physically covered by flood waters. Producing these scenarios required modeling the combined effects of sea level rise and changing peak flows along the relevant rivers, which in turn are derived from climate models and downscaling methods described above. Thus, we used a chain of computational models to incorporate all these effects as shown in the flow chart in Fig. 3. The climate data, storm surge, and sea-level rise inputs used depends on the climate change scenario and downscaling method selected. By adjusting these inputs, we generate multiple inundation results to provide a range of scenarios for planning.

Regional climate scenarios used in this study follow the methods described above, resulting in future projections of temperature, precipitation, and streamflow. To simulate the resulting changes in flood inundation in the lower Snohomish River, we coordinated with WEST Consultants, who developed a one-dimensional unsteady-flow hydraulic model for the lower Snohomish River using the US Army Corps of Engineers Hydrologic Engineering Center's River Analysis System

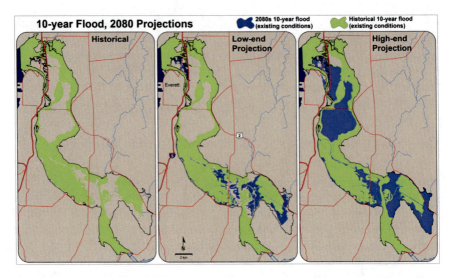

Fig. 3 10-year flood inundation simulated for historic conditions and projected 2080s climate. Green areas are inundated during historical 10-year flood events. Blue areas are flooded in addition during projected future 10-year events based on a low-end and high-end climate projection. *Source* authors own

(HEC-RAS) model. Projections of changing storm surge, SLR, and riverine flooding provided boundary conditions for the hydraulic model, which was used to estimate the combined effects of sea level rise and changing peak flows on flood inundation.

As an upstream boundary condition for simulating current climate conditions, we used existing hydrographs representing the historical 10- and 100-year flood events. These were developed for a Flood Insurance Study (FIS) for Federal Emergency Management Agency (FEMA) Region 10. We then developed projected future flood events by scaling these historical hydrographs. The scaling factors were determined using the ratio of simulated future to historical flow intensities at the corresponding return periods. The future flood events were simulated using the methods described in the previous section: temperature and precipitation are obtained from both the statistical downscaling and regional climate model (WRF) and streamflow from the VIC hydrologic model.

For flood statistics, we extracted the 1-, 3-, 5-, and 7-day consecutive highest flows for each year and ranked the values by flow magnitude. A quantile was assigned to each value using an unbiased quantile estimator (Maidment 1993). These were fit to a Generalized Extreme Value (GEV) Distribution using L-moments (Hosking and Wallis 1993; Wang 1997). For historical as well as two future time periods, we estimated flood magnitudes with 10-year (Q10), 50-year (Q50) and 100-year (Q100) return frequencies from the fitted GEV distributions. These are defined based on the probability that peak flows exceed a certain threshold on any given year. Sometimes referred to as the "Annual Chance of Exceedance" (ACE), the 3 return frequencies correspond to an ACE of 10, 2, and 1%, respectively.

The ratios of future to historical flood events at a given return interval are used to scale observationally-based hydrographs of historical extreme flows. The resulting maximum, median and minimum change projected for the Snohomish River by all ten GCMs is listed in Table 1. Note that there is some tendency towards larger changes for the longer period flows (e.g.: 7-day relative to 1-day), especially for the 50- and 100-year events, although the differences are small compared to the range among models.

The lower, marine boundary condition to the model is determined by total water levels associated with sea level rise, tides, and surge. It is possible that climate change may affect extremes in surface pressure, winds, or circulation due to changes in storm frequency and strength. Hamman et al. (2016) evaluated this possibility, using regional climate model simulations (Salathé et al. 2010, 2014) and a regression model trained on regional variations in sea level pressure and sea surface temperature associated with the El Niño Southern Oscillation (ENSO). Their results, consistent with the findings of Stammer and Hüttemann (2008),

Table 1 Ratio of future to historical floods for the Snohomish River for 1-day, 3-day, 5-day and 7-day consecutive highest flows with a return frequency of 10-year (Q10), 50-year (Q50) and 100-year (Q100) for the 2040s and 2080s. These correspond to the 10, 2, and 1% ACE, respectively. Citation needed!

Time periods	Return interval	Ratio of future to historical peak flow				
			1-Day	3-Day	5-Day	7-Day
2040s	Q10 (10%ACE)	Max	1.60	1.61	1.65	1.65
		Mean	1.25	1.27	1.29	1.29
		Min	1.06	1.08	1.09	1.09
	Q50 (2%ACE)	Max	1.49	1.47	1.51	1.54
		Mean	1.15	1.17	1.20	1.22
		Min	1.01	1.03	1.04	1.04
	Q100 (1%ACE)	Max	1.44	1.41	1.45	1.49
		Mean	1.10	1.12	1.16	1.18
		Min	0.96	0.97	1.03	1.02
2080s	Q10 (10%ACE)	Max	1.78	1.79	1.83	1.82
		Mean	1.40	1.42	1.43	1.43
		Min	1.16	1.18	1.20	1.20
	Q50 (2%ACE)	Max	1.64	1.63	1.64	1.66
		Mean	1.27	1.30	1.31	1.33
		Min	1.04	1.08	1.10	1.12
	Q100 (1%ACE)	Max	1.57	1.56	1.56	1.59
		Mean	1.22	1.25	1.26	1.28
		Min	0.98	1.04	1.06	1.08

suggest that climate change has very little influence on storm surge. Thus, we incorporated the effects of storm surge using the most recent 50-years of hourly observations from Seattle. From these data, we estimate the 10, 50, and 100-year surge values (relative to MHHW) using the same approach described above for peak flows.

Sea Level Rise (SLR) projections were taken from the recent synthesis of projections for the West Coast by the National Research Council (NRC 2012). Among projections of global sea level rise, the NRC projections are within the range of other recent projections—higher than the projections of the recent Intergovernmental Panel on Climate Change (IPCC 2013) report, but lower than those of Vermeer and Rahmstorf (2009). For consistency with the time periods used for the study, the sea level projections were interpolated to 2045 and 2085 using a quadratic fit. The resulting SLR projection for the 2040 s are 3.7 in (low) 5.3 in (medium) 7.3 in (high), and for the 2080 s 9.8 in (low) 16 in (medium) 22 in (high). Subsiding land motion in this location increases the actual sea-level rise, and we accounted for this effect using an estimate of −1 mm/yr for Anacortes, WA (NRC 2012).

In an additional set of simulations we evaluated the impact of levee modifications. In consultation with staff at Snohomish County, we identified two alternative levee scenarios which we could model. The first involved removing the levees protecting Crabb and Beck dikes, which are near the upstream end of the model domain, near the city of Monroe, Washington. The second alternative involved breaching the levees protecting Spencer Island, allowing for the flooding of the island and providing storage area for excess flow volumes that would cause flooding elsewhere. We do not present maps of the results of these simulations since neither had an appreciable effect on flooding due to the lack of adequate storage to accommodate flood water. Although there are other options for providing flood storage, few are currently viable as options given current floodplain development and land use constraints.

Results from this study were incorporated into TNC's "coastalresilience" decision support web tool, which allows users to interactively explore the study results, along with numerous other spatial datasets. Figures 3 and 4 show screenshots of inundation maps as they are displayed in TNC's web tool. These maps illustrate changes both in depth and area. Changes in the depth of inundation, in contrast to areal extent alone, are notable for all scenarios. Projected changes in the areal extent of flooding are large for the 10-year flood, but quite small for the 100-year flood. This is not surprising, since the levees in the lower basin are primarily designed to mitigate 10-year events. This means that the historical 100-year flood, under present-day conditions, should already result in flooding that extends from valley wall to valley wall, which limits the potential changes in flood extent going into the future. In contrast, small changes in the volume of the 10-year flood may lead to large changes in the area inundated. As a result, we expect the area inundated to increase more for moderate flood events rather than for the most extreme events.

A primary goal of this study was to provide a proof-of-concept for incorporating climate change into flood risk assessment and planning. By using a hydraulic model

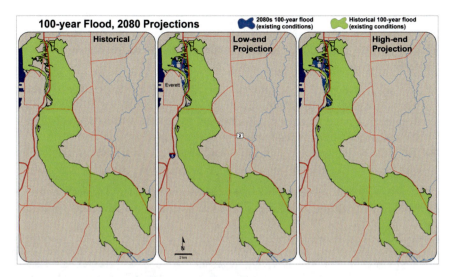

Fig. 4 As for Fig. 3, but for 100-year flood events *Source* authors own

that was essentially off-the-shelf, we were able to assess the combined impacts of SLR and streamflow on flooding at a relatively low cost. Having now established the methodology, this approach could be applied elsewhere in the region at a much lower cost.

Conclusions

The case study cited in this chapter focused on just three pathways for climate change impacts on floodplains: sea level rise, reduced snowpack and higher intensity precipitation extremes. These are key factors, but there are other mechanisms by which climate may impact flood risk such as vegetation loss due to wildfires or stream channel changes due to sediment transport and landslides. Climate change impacts on floodplains also extend beyond changes in flood risk discussed here but also include the impacts on riparian habitat, groundwater, saltwater intrusion, and water temperature. Thus, more work is needed to evaluate these risks and determine their relevance to managers, tribes, agriculture, and other key stakeholders.

Despite evidence that the response of extreme events to climate change is highly dependent on local processes that are not well represented in current global models, substantial fundamental questions remain unanswered. For example, regional climate models can be used to answer some of these questions, but the differences between global and regional simulations of extreme events have not been rigorously examined, and the suitability of regional climate models for specific applications is not well established. Important decisions with significant economic and societal implications will be made in the next few years based on our incomplete understanding of how climate change affects extreme events. In particular, we do not currently understand:

(1) The relative influences of climate variability and climate change on recent trends in heavy precipitation;
(2) The relative importance of large-scale and mesoscale processes on changes in the frequency, duration, and intensity of heavy precipitation; and
(3) The relative importance of precipitation, snowpack dynamics, and antecedent conditions in connecting climate change to flood risk.

These issues must be settled in order to better understand and project changes in extreme events a changing climate and to evaluate potential adaptation strategies.

Acknowledgements Research reported in this chapter has been supported by the Department of Homeland Security through the Critical Infrastructure Resilience Institute (CIRI) and by The Nature Conservancy (TNC). We thank Kris Johnson at TNC for developing the original idea for the Snohomish inundation study. We would also like to thank Ray Walton and Sarah Bengston at WEST consultants for their work developing and post-processing the hydraulic model output.

References

Allen MR, Ingram WJ (2002) Constraints on future changes in climate and the hydrologic cycle. Nature 419:224-232

Avise J, Chen J, Lamb B, Wiedinmyer C, Guenther A, Salathe E, Mass C (2009) Attribution of projected changes in summertime US ozone and PM2.5 concentrations to global changes. Atmos Chem Phys:1111–1124

Chang EKM (2007) Assessing the increasing trend in Northern hemisphere winter storm track activity using surface ship observations and a statistical storm track model. J Clim 20:5607–5628

Christensen JH, Boberg F, Christensen OB, Lucas-Picher P (2008) On the need for bias correction of regional climate change projections of temperature and precipitation. Geophys Res Lett 35: L20709

Colle BA, Mass CF (1996) An observational and modeling study of the interaction of low-level southwesterly flow with the Olympic Mountains during COAST IOP 4. Mon Weather Rev 124:2152–2175

Daly C, Neilson RP, Phillips DL (1994) A statistical topographic model for mapping climatological precipitation over mountainous terrain. J Appl Meteorol 33:140–158

DeGaetano AT (2009) Time-dependent changes in extreme-precipitation return-period amounts in the continental United States. J Appl Meteorol Climatol 48:2086–2099

Dettinger M (2011) Climate change, atmospheric rivers, and floods in California—a multimodel analysis of storm frequency and magnitude changes. JAWRA J Am Water Resour Assoc 47:514–523

Duliére V, Zhang Y, Salathé EP (2013) Changes in twentieth-century extreme temperature and precipitation over the Western United States based on observations and regional climate model simulations. J Clim 26:8556–8575

Easterling DR, Evans JL, Groisman PY, Karl TR, Kunkel KE, Ambenje P (2000) Observed variability and trends in extreme climate events: a brief review. Bull Am Meteor Soc 81: 417–425

Elsner MM, Cuo L, Voisin N, Deems JS, Hamlet AF, Vano JA, Mickelson KEB, Lee SY, Lettenmaier DP (2010) Implications of 21st century climate change for the hydrology of Washington State. Clim Change 102:225–260

Field CB, Barros V, Stocker TF, Qin D, Dokken DJ, Ebi KL, Mastrandrea MD, Mach KJ, Plattner G-K, Allen SK, Tignor M, Midgley PM (eds) (2011) IPCC, 2012: managing the risks of extreme events and disasters to advance climate change adaptation. A special report of working groups I and II of the Intergovernmental panel on climate change, Cambridge University Press, p 582

Fowler HJ, Blenkinsop S, Tebaldi C (2007) Linking climate change modelling to impacts studies: recent advances in downscaling techniques for hydrological modelling. Wiley, pp 1547–1578

Fowler HJ, Cooley D, Sain SR, Thurston M (2010) Detecting change in UK extreme precipitation using results from the climateprediction.net BBC climate change experiment. Extremes 13:241–267

Fowler HJ, Wilby RL (2010) Detecting changes in seasonal precipitation extremes using regional climate model projections: implications for managing fluvial flood risk. Water Resour Res 46

Frei C, Schär C (2001) Detection probability of trends in rare events: theory and application to heavy precipitation in the Alpine region. J Clim 14:1568–1584

Garvert MF, Smull B, Mass C (2007) Multiscale mountain waves influencing a major orographic precipitation event. J Atmos Sci 64:711–737

Groisman PY, Knight RW, Easterling DR, Karl TR, Hegerl GC, Razuvaev VAN (2005) Trends in intense precipitation in the climate record. J Clim 18:1326–1350

Hamlet AF, Elsner MM, Mauger GS, Lee S-Y, Tohver I, Norheim RA (2013) An overview of the columbia basin climate change scenarios project: approach, methods, and summary of key results. Atmos Ocean 51:392–415

Hamman JJ, Hamlet AF, Lee SY, Fuller R, Grossman EE (2016) Combined effects of projected sea level rise, storm surge, and peak river flows on water levels in the Skagit Floodplain. Northwest Sci 90:57–78

Hattermann FF, Huang SC, Burghoff O, Hoffmann P, Kundzewicz ZW (2016) Brief Communication: an update of the article modelling flood damages under climate change conditions—a case study for Germany. Nat Hazards Earth Syst Sci 16:1617–1622

Held IM, Soden BJ (2006) Robust responses of the hydrological cycle to global warming. J Clim 19:5686–5699

Hosking J, Wallis J (1993) Some statistics useful in regional frequency-analysis. Water Resour Res 29:271–281

IPCC (2007) Climate change 2007: the physical science basis. Contribution of Working Group I to the Fourth assessment report of the Intergovernmental panel on climate change. Cambridge University Press, Cambridge, United Kingdom and New York, NY, USA

IPCC (2013) Climate change 2013: the physical science basis. Contribution of Working Group I to the Fifth assessment report of the Intergovernmental panel on climate change. Cambridge University Press

Leung LR, Qian Y (2009) Atmospheric rivers induced heavy precipitation and flooding in the western U.S. simulated by the WRF regional climate model. Geophys Res Lett 36

Liang X, Lettenmaier DP, Wood EF, Burges SJ (1994) A simple hydrologically based model of land-surface water and energy fluxes for general-circulation models. J Geophys Res-Atmos 99:14415–14428

Lohmann D, NolteHolube R, Raschke E (1996) A large-scale horizontal routing model to be coupled to land surface parametrization schemes. Tellus Ser A-Dyn Meteorol Oceanogr 48:708–721

Maidment DR (1993) Handbook of hydrology. McGraw-Hill, New York

Mass C, Skalenakis A, Warner M (2011) Extreme precipitation over the West Coast of North America: is there a trend? J Hydrometeorol 12:310–318

Meehl GA, Covey C, Delworth T, Latif M, McAvaney B, Mitchell JFB, Stouffer RJ, Taylor KE (2007) The WCRP CMIP3 multimodel dataset—a new era in climate change research. Bull Am Meteor Soc 88:1383–1394

Min S-K, Zhang X, Zwiers FW, Hegerl GC (2011) Human contribution to more-intense precipitation extremes. Nature 470:378–381

Mote PW, Salathe EP (2010) Future climate in the Pacific Northwest. Clim Change 102:29–50

Neiman PJ, Ralph FM, Wick GA, Kuo YH, Wee TK, Ma ZZ, Taylor GH, Dettinger MD (2008a) Diagnosis of an intense atmospheric river impacting the Pacific Northwest: storm summary and offshore vertical structure observed with COSMIC satellite retrievals. Mon Weather Rev 136:4398–4420

Neiman PJ, Ralph FM, Wick GA, Lundquist JD, Dettinger MD (2008b) Meteorological characteristics and overland precipitation impacts of atmospheric rivers affecting the West Coast of North America based on eight years of SSM/I satellite observations. J Hydrometeorol 9:22–47

Neiman PJ, Schick LJ, Ralph FM, Hughes M, Wick GA (2011) Flooding in Western Washington: the connection to atmospheric rivers. J Hydrometeorol 12:1337–1358

NRC (2012) Sea-level rise for the coasts of California, Oregon, and Washington: past, present, and future. The National Academies Press, Washington, DC

Pall P, Allen MR, Stone DA (2007) Testing the Clausius-Clapeyron constraint on changes in extreme precipitation under CO2 warming. Clim Dyn 28:351–363

Rupp DE, Abatzoglou JT, Hegewisch KC, Mote PW (2013) Evaluation of CMIP5 20th century climate simulations for the Pacific Northwest USA. J Geophys Res-Atmos 118:10884–10906

Rybczyk JM, Hamlet AF, MacIlroy C, Wasserman L (2016) Introduction to the Skagit issue from glaciers to estuary: assessing climate change impacts on the Skagit River Basin. Northwest Sci 90:1–4

Salathé E (2006) Influences of a shift in North Pacific storm tracks on western North American precipitation under global warming. Geophys Res Lett

Salathé EP, Hamlet AF, Mass CF, Lee SY, Stumbaugh M, Steed R (2014) Estimates of twenty-first-century flood risk in the Pacific Northwest based on regional climate model simulations. J Hydrometeorol 15:1881–1899

Salathé EP, Leung LR, Qian Y, Zhang YX (2010) Regional climate model projections for the State of Washington. Clim Change 102:51–75

Salathé EP, Steed R, Mass CF, Zahn P (2008) A high-resolution climate model for the U.S. Pacific Northwest: mesoscale feedbacks and local responses to climate change. J Clim 21:5708–5726

Stammer D, Hüttemann S (2008) Response of regional sea level to atmospheric pressure loading in a climate change scenario. J Clim 21:2093–2101

Taylor KE, Stouffer RJ, Meehl GA (2011) An overview of CMIP5 and the experiment design. Bull Am Meteor Soc 93:485–498

Tebaldi C, Hayhoe K, Arblaster JM, Meehl GA (2006) Going to the extremes. Clim Change 79:185–211

Tohver IM, Hamlet AF, Lee S-Y (2014) Impacts of 21st-century climate change on hydrologic extreme in the Pacific Northwest region of North America. JAWRA J Am Water Resour Assoc:n/a-n/a

Trenberth KE (2011) Changes in precipitation with climate change. Climate Res 47:123–138

Ulbrich U, Pinto JG, Kupfer H, Leckebusch GC, Spangehl T, Reyers M (2008) Changing Northern Hemisphere storm tracks in an ensemble of IPCC climate change simulations. J Clim 21:1669–1679

Vermeer M, Rahmstorf S (2009) Global sea level linked to global temperature. Proc Natl Acad Sci USA 106:21527–21532

Wang Q (1997) LH moments for statistical analysis of extreme events. Water Resour Res 33:2841–2848

Warner MD, Mass CF, Salathe EP (2012) Wintertime extreme precipitation events along the Pacific Northwest Coast: climatology and synoptic evolution. Mon Weather Rev 140:2021–2043

Warner MD, Mass CF, Salathe EP (2015) Changes in winter atmospheric rivers along the North American West coast in CMIP5 climate models. J Hydrometeorol 16:118–128

White AB, Colman B, Carter GM, Ralph FM, Webb RS, Brandon DG, King CW, Neiman PJ, Gottas DJ, Jankov I, Brill KF, Zhu Y, Cook K, Buehner HE, Opitz H, Reynolds DW, Schick LJ (2011) NOAA's rapid response to the Howard A. Hanson dam flood risk management crisis. Bull Am Meteor Soc 93:189–207

Wood AW, Leung LR, Sridhar V, Lettenmaier DP (2004) Hydrologic implications of dynamical and statistical approaches to downscaling climate model outputs. Clim Change 62:189–216

Zhang Y, Duliere V, Mote PW, Salathé EP (2009) Evaluation of WRF and HadRM mesoscale climate simulations over the U.S. Pacific Northwest. J Clim 22:5511–5526

Chapter 8
The Impact of Climate Change on Resilience of Communities Vulnerable to Riverine Flooding

Xianwu Xue, Naiyu Wang, Bruce R. Ellingwood and Ke Zhang

Abstract Riverine flooding due to intense precipitation or snowmelt is one of the most devastating natural hazards in the United States in terms of annual damages and economic losses to the built environment and social impacts on communities. Flood inundation mapping, where the likely depths of extreme floods are placed on a map of the community, is important for evaluating flood risks and for enhancing community resilience. However, the Flood Insurance Rate Maps developed by the Federal Emergency Management Agency are not adequate for the evolving needs for community resilience assessment and decision-making over the next century, during which climate change effects are likely to be significant. In this study, we develop a flood hazard modeling framework to support community resilience assessment. This framework couples a hydrological model, which simulates the hydrological processes in a community at a coarser resolution using measured and/or remote sensed precipitation, with a hydraulic analysis module, which computes localized flood depths, velocities and inundated areas at a finer spatial resolution. The Wolf River Basin in Shelby County, Tennessee, which includes the city of

---

X. Xue · N. Wang
School of Civil Engineering and Environmental Sciences,
University of Oklahoma, Norman, OK 73019, USA
e-mail: xuexianwu@ou.edu

N. Wang
e-mail: naiyu.wang@ou.edu

B. R. Ellingwood (✉)
Department of Civil and Environmental Engineering,
Colorado State University, Ft. Collins, CO 80523, USA
e-mail: bruce.ellingwood@colostate.edu

K. Zhang
Cooperative Institute for Mesoscale Meteorological Studies, The University of
Oklahoma, 120 David L. Boren Blvd, Norman, OK 73072, USA
e-mail: kezhang@ou.edu

K. Zhang
And State Key Laboratory of Hydrology-Water Resources and Hydraulic
Engineering, Hohai University, 1 Xikang Road, Nanjing,
Jiangsu Province 210098, China

Memphis, is used as a testbed to calibrate and validate this coupled model using precipitation and streamflow data obtained from gauge stations operated by the US Geological Survey and to illustrate the potential impacts of climate change in the 21st Century on civil infrastructure, revealing that such impacts are non-negligible but are manageable by proper engineering.

Introduction

Floods are among the most common and devastating natural hazards in the United States based on the U.S. Natural Hazard Statistics for 2015 (National Weather-Service 2016). All 50 states have experienced floods or flash floods during the past five years according to the Federal Emergency Management Agency (FEMA) (FEMA 2016a). Floods are a threat to human life as well as a source of property damage. The total annual flood insurance claims in the United States average more than $3.5 billion (FEMA 2016a; Molk 2016). The potential exists for even larger losses in the future, given the shifts of population to hazard-prone areas of the United States and global climate change. Moreover, floods are often coupled with the occurrence of other natural hazards such as heavy rainfall, hurricanes and tornados, amplifying their risk and impact. The resilience of a community to such natural disasters is reflected in the ability of its physical infrastructure and socioeconomic institutions to return to a level of normalcy within a reasonable time following the occurrence of an event (PPD-21 2013).

Flood inundation mapping is important for evaluating community flood risks as well as for enhancing societal resilience to flood hazards. Traditionally, flood risk at community or regional scales has been assessed using the Flood Insurance Rate Maps (FIRMs) and Flood Insurance Study (FIS) reports developed by the Federal Emergency Management Agency (FEMA) for the National Flood Insurance Program (NFIP) (FEMA 2016b) from statistical data analyses of river flow, storm tides, rainfall, hydrologic/hydraulic analyses and topographic surveys. However, the FIS and FIRM have the following limitations for assessing community resilience to flood hazard:

1. Many of the studies utilized in the FIRMs and FIS reports are often more than ten years (or more) old and may not reflect recent patterns of urbanization that might affect the current flood risk (Xian et al. 2015).
2. The FIRMs and FIS reports depict the 100-year (Base Flood Elevation, or BFE) and 500-year return period flood contours. They do not reflect specific flood scenarios. Scenario events are a common basis for community resilience planning because they need not be associated with a specific probability of being exceeded,[1] and are needed for characterizing the correct spatial distribution of risk.

---

[1]*Return period events* associated with spatially distributed hazards represent an aggregation of events. For example, the 2500-*year return period earthquake* for Shelby County, TN has a 5%

3. The FIRMs only provide information on flood inundation area and depth, and cannot directly support damage assessment and loss estimation to the built environment, which depend on both water depth and flow velocity (Karamouz et al. 2016).
4. The FIRMS do not reflect potential impacts of climate change (Zahmatkesh et al. 2015).

The assessment of community resilience under flooding events requires new and innovative methods for modeling flood hazard. A combination of hydrologic and hydraulic models is required to determine risks due to flooding. Hydrological models are focused on the distribution of water over, beneath the earth surface and through the atmosphere, and focus on transforming precipitation into quantity of runoff. In contrast, hydraulic models utilize principles of fluid mechanics to simulate the movement of water beneath and over the earth's surface or in river channels. While hydraulic models are theoretically the most advanced models for producing flood elevations, velocities and flood plain contours (Aggett and Wilson 2009; Zhu et al. 2016), most hydraulic models require inflow data either from gauge measurements or from the hydrological simulation. Since coupled hydraulic-hydrological models to simulate flood inundations along the river channels can be time-consuming for a large basin (Hagen et al. 2010), a balance between model complexity and computational efficiency is needed in practice (Abatzoglou 2013; Zhao and Shao 2015).

Accordingly, in this study, we develop a simple integrated modeling framework that couples a hydrological model, which simulates the hydrological processes in a region at a coarser resolution using measured and/or remotely sensed precipitation data, with a hydraulic analysis module, which computes localized flood depths, velocities and inundated areas at a finer spatial resolution. Following its validation using existing streamflow gauge data, we use this model to forecast the potential impacts of climate change in the 21st Century on flood hazards and their impact on civil infrastructure in Shelby County, TN.

Coupled Hydrological/Hydraulic Flood Hazard Model

Figure 8.1 summarizes the framework to integrate hydrological modeling with hydraulic inundation modeling (hereafter denoted as iH&H) through one-way coupling for flood simulation and inundation mapping in ungauged areas and

---

damped spectral acceleration at a period of 0.2 s that is approximately 0.9 g (883 cm/s^2) at various rock sites in the County. This value can result from any one of several possible earthquakes occurring in the New Madrid Seismic Hazard Zone. A *scenario earthquake* for Shelby County, TN would involve selecting one of these possible earthquakes (e.g., a magnitude 7.7 earthquake with an epicenter 50 km from downtown Memphis, and determining the resulting spatial distribution of ground motions and their impact on the physical infrastructure of the community. The problem is similar for flood hazard assessment.

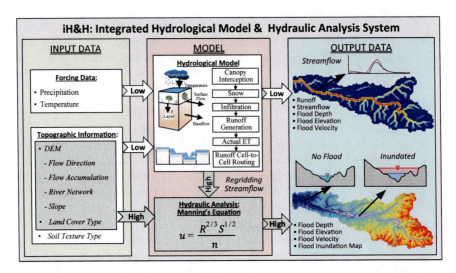

Fig. 8.1 Integrated hydrological model and hydraulic analysis system for predicting streamflow and flood depth, elevation and velocity, and inundation at different scales of resolution

macro-scale regions. The left panel of Fig. 8.1 shows that the forcing data consist of only precipitation and air temperature, while the topographic data include the digital elevation model (DEM) describing changes in elevation, land cover type and soil texture type. The other topographic data, such as flow direction, flow accumulation, river network, and slope, can be derived from the DEM.

Grid-Based Distributed Hydrological Model

Any grid-based distributed hydrological model can be used in the iH&H system to simulate streamflow in a river basin at a relatively coarse spatial resolution. In this study, we utilize a coupled routing and excess storage (CREST) hydrological model jointly developed by the University of Oklahoma (http://hydro.ou.edu) and the NASA SERVIR Project Team (www.servir.net) (Wang et al. 2011; Xue et al. 2013). The SCE-UA method (shuffled complex evolution method developed by the University of Arizona) (Duan et al. 1992) is implemented to calibrate the CREST model parameters (Xue et al. 2013). Daily potential evapotranspiration (PET) is incorporated in CREST using the model developed by Hamon (1961). This grid-based distributed hydrological model can provide high-resolution streamflow input for hydraulic analysis (discussed subsequently in 2.2) to produce localized flood depths, velocities and inundation maps (right panel of Fig. 8.1, showing the study area described in Sect. 8.3). The terms Low and High in Fig. 8.1 identify those portions of the analysis that are performed at low resolution (e.g., precipitation or temperature; hydrologic modeling) or at high resolution (hydraulic analysis).

Hydraulics-Based Inundation Mapping Model

Advanced hydraulic models require high-resolution river cross-sectional geometry data and detailed initial and boundary conditions. In practice, these data are difficult to obtain in remote regions or areas that are larger than 10,000 km² (3861 mi²). Moreover, advanced hydraulic models are computationally intensive, making them hard to apply except on a local scale (Hagen et al. 2010). Alternatively, a simplified hydraulic model can be used initially to map the areas inundated by floods. Given a high resolution DEM, we can derive the cross-section profile of each river segment. The discharge, Q, at a cross-section of the river is computed by dividing the cross-section into n equal-width elements (Genc et al. 2015),

$$Q = \sum_{i=1}^{n} u_i d_i \Delta w_i \tag{8.1}$$

where u_i, d_i and Δw_i are the mean velocity, mean flood depth and width in cross sectional element, i, respectively. The flow velocity can be estimated from Manning's equation (Chow et al. 1988):

$$u_i = \frac{R_i^{2/3} S_i^{1/2}}{n_i} \tag{8.2}$$

where $R_i = A_i/P_i$, is the hydraulic radius of the cross section i of the channel, A_i is the cross section area, P_i is the wetted perimeter; S_i is the friction slope, which equals the bed slope for a steady uniform flow; n is Manning's coefficient. Substituting Eq. (8.2) into Eq. (8.1), Q becomes:

$$Q = \sum_{i=1}^{n} \frac{R_i^{2/3} S_i^{1/2}}{n_i} d_i \Delta w_i = \sum_{i=1}^{n} \frac{d_i^{5/3} S_i^{1/2} \Delta w_i}{n_i} \tag{8.3}$$

The discharge, Q, in Eq. (8.3) simulated by the hydrological model at a coarse resolution then can be transformed to a finer-resolution streamflow based on the river channel and the DEM.

As will be seen subsequently, the iH&H system has the following three advantages: (1) the coupled approach is sufficiently efficient to capture the characteristics of the flood events; (2) the physics-based approach enables long-term forecasting of future extreme flood events projected by Intergovernmental Panel on Climate Change (IPCC) climate scenarios; and (3) the model outputs, i.e. flood depths, velocities and inundation maps, directly support scenario-based infrastructure loss estimation in community resilience assessment.

Assessment of Flood Hazard for Shelby County, TN

Description of the Wolf River Basin

Shelby County, TN is situated in the southwest corner of the State of Tennessee with the Mississippi River bordering it on the west. It includes the city of Memphis and is that state's largest county both in terms of population and geographic area. The Wolf River basin, with an area of approximately 1432 km^2, drains a large portion of the northern and eastern areas of Shelby County. The Wolf River stretches 165 km from the Holly Springs National Forest to the East to the Mississippi River on the West (Fig. 8.2). Elevations range from 62.75 m at the Mississippi River outlet to 235.19 m at the headwater origin to the west (Fig. 8.2). The various vegetation types within this basin based on National Land Cover Database (Homer 2015), along with the Manning coefficients for these vegetation types (Mattocks and Forbes 2008) are summarized in Table 8.1.

The IH&H System and Supporting Databases

The hydrological model uses a spatial resolution of 1 km (30 arc-seconds) and a temporal resolution of 1 day. The low-resolution (30 arc-second) DEM and data on flow direction and accumulation were obtained from the United States Geological Survey (USGS) geo-referenced datasets (http://hydrosheds.cr.usgs.gov/). The CREST model uses these data to establish topological connections among grid cells and to derive other topographical data, such as river network and slope. Data on daily precipitation and minimum and maximum temperature for 1979–2014 were obtained from the University of Idaho Gridded Surface Meteorological Database (METDATA, http://metdata.northwestknowledge.net/). The METDATA were then interpolated to 30 arc-seconds spatial resolution to match the spatial resolution of the DEM and

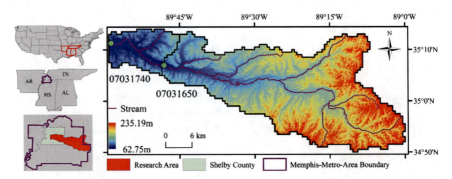

Fig. 8.2 Study domain showing the boundary of the Wolf River Basin in southwest TN with its river network, the locations of two streamflow gauges

Table 8.1 NLCD land cover classes, their Manning coefficients and their percent areas percentage in the Wolf River Basin

NLCD Class	Description	Percentage (%)	Manning n
11	Open water	0.91	0.020
21	Developed, Open space	8.53	0.020
22	Developed, Low intensity	6.20	0.050
23	Developed, Medium intensity	3.67	0.100
24	Developed, High intensity	1.01	0.150
31	Barren land (Rock/Sand/Clay)	0.03	0.090
41	Deciduous forest	20.40	0.100
42	Evergreen forest	4.21	0.110
43	Mixed forest	2.79	0.100
52	Shrub/scrub	12.80	0.050
71	Grassland/herbaceous	1.07	0.034
81	Pasture/hay	13.69	0.033
82	Cultivate crops	15.75	0.037
90	Woody wetlands	8.54	0.100
95	Emergent herbaceous wetlands	0.41	0.045

hydrological simulation. The high-resolution (1 arc-second, corresponding to approximately 30 m^2) DEM for the inundation mapping was obtained from the National Elevation Dataset (NED), which is the USGS's primary elevation data product (Gesch et al. 2009).

USGS streamflow Gauge #07031740, located on the Wolf River nearest to the outlet (Fig. 8.2), is treated as the outlet of the Wolf River Basin. The observed streamflow data for Gauge #07031740 is available from January 1979 to December 2014 (36 years). USGS streamflow Gauge #07031650 at Germantown, TN, located at the upper boundary of the urban area (Fig. 8.2), is used as a validation station. The observed streamflow data for Gauge #07031650 is available from January 1979 to December 1986 (8 years) and then from January 1990 to December 2014 (25 years).

Model Calibration and Validation

It is impossible to characterize all spatial and temporal variabilities on a watershed scale, and some parameters in the model must be calibrated to streamflow data (Refsgaard 1997). Accordingly, the iH&H framework was *calibrated* for the 3-yr period from January 2001 to December 2003 using the automatic calibration method (SCE-UA—Duan et al. 1992) for simulated and observed daily streamflow at Station #07031740 using the Nash-Sutcliffe (1970) coefficient of efficiency (NSCE), a common method for assessing the predictive power of a hydrological model.

In contrast to calibration, *validation* is the process of demonstrating that a given site-specific model is capable of making sufficiently accurate predictions. Following calibration, the model was validated using the calibrated parameters above for the data collected at Stations #07031740 and #07031650 for the periods from January 1979 to December 2000 and from January 2004 to December 2014. Figure 8.3 compares the simulated (blue) and observed (red) daily streamflow at USGS stations #07031740 (Fig. 8.3a) and #07031650 (Fig. 8.3b), along with daily precipitation (black). The comparison between predictions and observations at station #07031740 shows good agreement for both the calibration period and the two validation periods. Similar results are found at station #07031650, even though the model was calibrated only at station #07031740, indicating that the iH&H model performs well in both the upstream and downstream stations of the urban area. Note that the model slightly underestimates the peaks at both stations, especially the extreme peaks at station #07031740. This may be due to human activities that are not captured in the model. However, these comparisons show that the iH&H framework is capable of predicting general trends in hydrological phenomena in Shelby County.

Probabilistic Analysis—Flood Return Periods

Probability-based estimates of intensities of hydrological parameters (such as streamflow or precipitation) during flood events are necessary for risk-informed assessment and decisions regarding civil infrastructure. Two common probability distributions, the Gumbel distribution and Log Pearson Type III distribution, are often used to develop the return period values of rainfall intensity, streamflow and flood depth from collected or simulated statistical data (Elsebaie 2012). In this study, we select the Gumbel distribution to determine the return period values[2] of interest for each flood parameter in each grid cell in the study area.

Figure 8.4 shows maps of precipitation, streamflow and flood depth with 50, 200, 500 and 1000-year return periods, respectively. Note that the flood depth (mm/day) in the right-hand figures is the overland flood depth, excluding the water depth in the river. As the return period of precipitation (storm) increases from 50 to 1000 years, the upstream area of the Wolf River Basin experiences larger increases in rainfall intensities (Fig. 8.4a, d, g, j). In contrast, streamflow (Fig. 8.4b, e, h, k) in the downstream area (downtown) increases more rapidly than in the middle to upper areas, suggesting that the residents in these areas will be exposed to higher potential flood risk. Figure 8.4c, f, i, l show the flood depth for the different return periods. The flood depth increases in the upstream area in a manner that is

---

[2]*The N-year return period value* of a parameter is that value with probability 1/N of being exceeded in any year. For example, the 100-year flood elevation is that elevation with a probability 1% of being exceeded in any year.

8 The Impact of Climate Change on Resilience of Communities ... 137

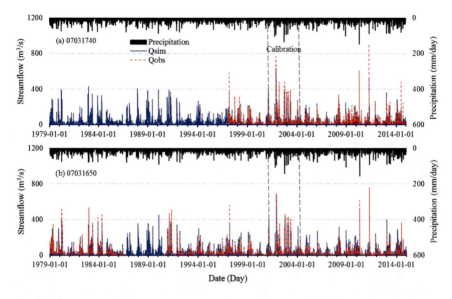

Fig. 8.3 Comparison of simulated daily streamflow forced by the METDATA precipitation and air mean temperature and observed streamflow in both calibration and two validation periods: **a** at station #07031740, **b** at station #07031650

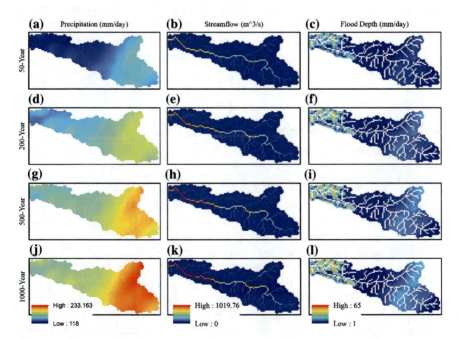

Fig. 8.4 Return period values for precipitation, streamflow, and overland flood depth (50, 200, 500 and 1000 years)

consistent with the precipitation changes. However, flood depth in the downstream area, including urban areas of Memphis, increases more significantly than in rural areas upstream due to a number of factors, including flatter topography and urbanization which decreases infiltration into the soil.

Climate Change Impacts on Flood Hazard in Shelby County, TN

Climate Change Projections and Downscaling

Climate change during the 21st Century may increase the susceptibility of Shelby County, TN to flood events. The possible effects of climate change on flood depth, inundation area, and flood velocity, and the impact on civil infrastructure, can be estimated using the coupled hydrologic/hydraulic models that were validated in the previous section. Fourteen global climate models (GCMs) from the Coupled Model Intercomparison Project Phase 5 (CMIP5) (Taylor et al. 2012) were used to assess the impact of climate changes on the streamflow of the Wolf River Basin under three projected Respective Concentration Pathways scenarios (RCP2.6, RCP4.5 and RCP8.5). Since these GCMs are produced from large-scale climate studies, they must be downscaled to finer spatial resolutions for use in the local analyses herein. We utilized the downscaled CMIP5 climate and hydrology projections for Shelby, TN from http://gdo-dcp.ucllnl.org/downscaled_cmip_projections/dcpInterface.html. To match the spatial resolution of the iH&H system, all the downscaled GCM forcings (daily precipitation, maximum and minimum temperatures) were interpolated to 30 arc-second grids.

Forecasts of Temperature and Precipitation for Wolf River Basin

Figure 8.5a shows the projected changes in Shelby County in yearly maximum temperature for each RCP for the current century. The temperature is projected to rise by 0.4–2.6 °C over the whole basin relative the present by 2099. The maximum daily temperature increases even more dramatically when compared to the ensemble mean daily temperature. Relative to the significant upward trends of maximum temperatures, the projections of precipitation in Fig. 8.5b are more uncertain. However, a closer look at the monthly average of daily maximum precipitation, as shown in Fig. 8.6a–c in three future epochs, demonstrates an increasing trend in the 21st Century. In particular, the maximum daily precipitation increases significantly in the February—May season, indicating that spring rainstorms are likely to become more severe toward the end of the century. Figure 8.5c

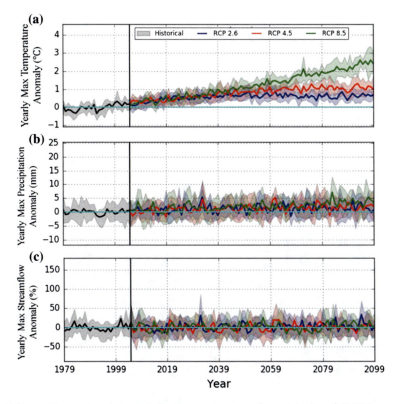

Fig. 8.5 a–c Time series of **a** annual maximum temperature, **b** precipitation of CMIP5 ensemble BCCA downscaling to Wolf River Basin and **c** streamflow of CMIP5 ensemble USGS Station #07031740 (Shadow shows 5–95% percentile)

shows that the projected yearly maximum streamflow increases under all RCPs, indicating that more severe floods are likely in the future. Sources of uncertainty in these figures reflect the uncertainty in ensemble precipitation and temperature from the 14 GCM models for three RCP scenarios used to predict the ensemble stream flow. The uncertainties in precipitation and temperature increase in the latter years of the 21st century, a trend that is especially pronounced for temperature as the 5–95% percentile bound (shadow) becomes wider towards 2099, as shown in Fig. 8.5a.

Flood Hazards During the 21st Century

To examine potential flood hazards in the 21st Century, we determined the expected maximum flood depth to occur during the next 100 years (from 2000 to 2099) for each of the grid cells in the hydrology model under each RCP scenario and using

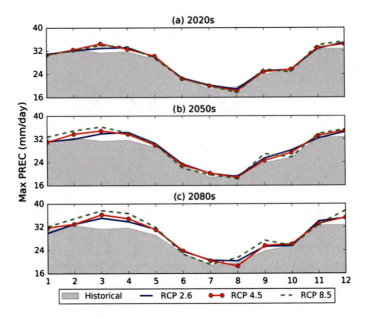

Fig. 8.6 Comparison of Monthly Average of maximum daily precipitation between historical trend (1980–2005, Gray Shadow) and future trajectories (2010–2099, lines)

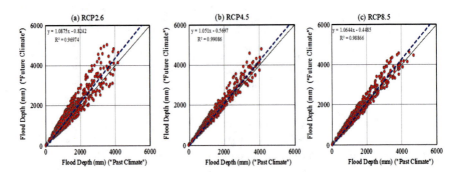

Fig. 8.7 a–c Comparison of 100-year return period flood-depth vs 100-year maximum flood depth forced by **a** RCP2.6, **b** RCP4.5 and **c** RCP8.5 scenarios from 2000 to 2099

each of the 14 GCMs. Figure 8.7a, b, and c compares this *expected* 100-*year maximum* flood depth (reflecting "future climate") with the 100-*year return period* flood depth (representing "past climate", in which the annual extremes are modeled as a stationary sequence described by a Gumbel distribution as discussed in Sect. 8.3.3) for all grid-cells. This comparison indicates that the "future" 100-*year maximum* flood depth exceeds the "historical" 100-*year return period* flood depth for a significant number of cells and GCMs in all RCP scenarios. While the 100-year maximum flood depth may be larger than or less than the 100-year return period flood depth (see Fig. 8.7) in certain locations (cells), the mean trend of the 100-year maximum flood depth over the entirety of Shelby County (dashed line) is invariably larger than the 100-year return period flood depth for all three climate change scenarios. Moreover, there are likely to be more extreme flood events that generate larger flood depths than the 100-year return period value reflected in the FIRMs. Accordingly, civil infrastructure facilities, e.g. dams, levees and buildings in flood plains, are likely to be susceptible to higher flood risk. It would be prudent for public planners and decision makers to institute policies to protect the residents of Shelby County from flood hazards, especially in its urban areas.

Climate Change Impact on Flood Protection Structures

Flood protection structures are found along the inland waterways in the United States. A typical flood wall is illustrated in Fig. 8.8. These cantilevered walls normally are reinforced concrete and are designed for a service period of 100 years. The dimensions of the wall vary, of course, but a typical wall might be 4.6–6.1 m in height and 457–586 mm in thickness. Concrete strength in the wall is typically is 27.6 MPa, and deformed bar reinforcement is typically ASTM A615 Grade 60 with yield strength 414 MPa.

The governing limit states for the design of these walls are strength in wall (stem) strength in flexure and in shear (USACE EM 1110-2-2104 2016). The design criteria for hydrostatic conditions are:

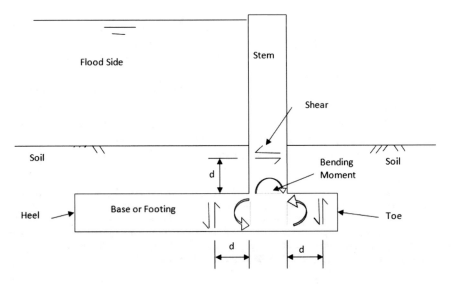

Fig. 8.8 Typical flood protection structure

$$\text{For flexure, } 0.90 M_n > 1.2\gamma H^3/6 \qquad (8.4)$$

$$\text{For shear, } 0.75 V_n > 1.2\gamma H^2/2 \qquad (8.5)$$

in which M_n and V_n are nominal strengths in flexure and in shear, H is the elevation of the water, and γ is the unit weight of water. A 10% increase in flood height due to climate change would cause a 33% increase in required flexural capacity and a 21% increase in required shear capacity to maintain approximately the same level of safety. These changes are manageable from a structural engineering point of view. For example, for a wall that is 457 mm (18 in) thick, with 51 mm (2 in) of bar cover, and bars spaced at 152 mm (6 in), increasing the reinforcement from No. 7 bars to No. 8 bars would increase the flexural capacity of the wall by approximately 29%.

Conclusions

This study presented a new framework that coupled a hydrological model to a hydraulic analysis system to investigate the impacts of riverine flooding from future storms on urban infrastructure situated along major rivers or in river basins. This iH&H framework performed well when compared to hydrologic data furnished by the USGS observing stations. The study also presented the impacts of climate change on the streamflow and flood depth through the next 100 years (from 2000 to 2099) in the Wolf Creek Basin in Shelby County, TN using three scenarios (RCP2.6, RCP4.5 and RCP8.5) identified in the latest IPCC study. The following conclusions may be drawn:

- Based on the climate change projections in the Wolf River Basin, temperature and precipitation are likely to increase relative to the values from 1980 to 2005, and the extreme flood events are likely to be more severe in the 21st century, particularly in the early spring months.
- The comparison of the 100-*year maximum* flood depth (for the future climate) with the 100-*year return period* flood depth (based on past climate) reveals that the intensity of extreme flooding is likely to increase due to climate change.
- In terms of the hydraulic forces on flood protection structures, the effect of climate change is manageable through sound engineering practices.

Acknowledgements The research reported herein was supported, in part, by the Center for Risk-Based Community Resilience Planning supported by the National Institute of Standards and Technology (NIST) under Cooperative Agreement No. 70NANB15H044. This support is gratefully acknowledged. Any opinions, findings and conclusions or recommendations expressed in this paper are those of the authors and do not necessarily reflect the views of the National Institute of Standards and Technology or the US Department of Commerce.

References

Abatzoglou JT (2013) Development of gridded surface meteorological data for ecological applications and modelling. Int J Climatol 33:121–131. https://doi.org/10.1002/joc.3413

Aggett GR, Wilson JP (2009) Creating and coupling a high-resolution DTM with a 1-D hydraulic model in a GIS for scenario-based assessment of avulsion hazard in a gravel-bed river. Geomorphology 113:21–34. https://doi.org/10.1016/j.geomorph.2009.06.034

Chow VT, Maidment DR, Mays LW (1988) Applied Hydrology. McGraw-Hill series in water resources

Duan Q, Sorooshian S, Gupta V (1992) Effective and efficient global optimization for conceptual rainfall-runoff models. Water Resour Res 28:1015–1031. https://doi.org/10.1029/91wr02985

Elsebaie IH (2012) Developing rainfall intensity–duration–frequency relationship for two regions in Saudi Arabia Journal of King Saud University. Eng Sci 24:131–140. https://doi.org/10.1016/j.jksues.2011.06.001

FEMA (2016a) Federal Emergency Management Agency Flood Facts. Retrieved from the Official Site of the NFIP. https://www.floodsmart.gov/floodsmart/pages/flood_facts.jsp

FEMA (2016b) Federal Emergency Management Agency Understanding Flood Maps. Retrieved from the Official Site of the NFIP. https://www.floodsmart.gov/floodsmart/pages/flooding_flood_risks/understanding_flood_maps.jsp

Genç O, Ardıçlıoğlu M, Ağıralioğlu N (2015) Calculation of mean velocity and discharge using water surface velocity in small streams. Flow Meas Instrum 41:115–120

Gesch D, Evans G, Mauck J, Hutchinson J, Jr. WJC (2009) The National Map—Elevation: U.S. Geological Survey Fact Sheet 2009–3053. U.S. Geological Survey

Hagen E, Shroder JF Jr, Lu XX, Teufert JF (2010) Reverse engineered flood hazard mapping in Afghanistan: a parsimonious flood map model for developing countries. Quatern Int 226:82–91. https://doi.org/10.1016/j.quaint.2009.11.021

Hamon WR (1961) Estimating potential evapotranspiration Journal of Hydraulics Division. Proc Am Soc Civ Eng 87:107–120

Homer C, Dewitz Jon, Yang L, Jin S, Danielson P, Xian G, Coulston J, Herold N, Wickham J, Megown K (2015) Completion of the 2011 National Land Cover Database for the conterminous United States-Representing a decade of land cover change information. Photogram Eng Remote Sens 81:345–354

Karamouz M, Fereshtehpour M, Ahmadvand F, Zahmatkesh Z (2016) Coastal flood damage estimator: an alternative to FEMA's HAZUS platform. J Irri Drainage Eng 142(6):04016016. https://doi.org/10.1061/(asce)ir.1943-4774.0001017

Mattocks C, Forbes C (2008) A real-time, event-triggered storm surge forecasting system for the state of North Carolina. Ocean Model 25:95–119. https://doi.org/10.1016/j.ocemod.2008.06.008

Molk P (2016) Private versus public insurance for natural hazards: individual behavior's role in loss mitigation. In: Gardoni P, Murphy C, Rowell A (eds) Societal risk management of natural hazards. Springer, Berlin, Germany

Nash JE, Sutcliffe JV (1970) River flow forecasting through conceptual models part I—a discussion of principles. J Hydrol 10:282–290

National Weather Service (2016) The Natural Hazard Statistics for 2015 in the United States. http://www.nws.noaa.gov/om/hazstats.shtml. Accessed 24 Mar 2016

PPD-21 (2013) Presidential policy directive—critical infrastructure security and resilience. The White House, Office of the Press Secretary. https://www.whitehouse.gov/the-press-office/2013/02/12/presidential-policy-directive-critical-infrastructure-security-and-resil. Accessed 17 Mar 2016

Refsgaard JC (1997) Parameterisation, calibration and validation of distributed hydrological models. J Hydrol 198(1–4):69–97

USACE (2016) Strength design for reinforced concrete hyudraulic structures. Engineer Manual 1110-2-2104, US Army Corps of Engineers, Washington, DC 20314

Taylor KE, Stouffer RJ, Meehl GA (2012) An overview of CMIP5 and the experiment design. Bull Am Meteor Soc 93:485–498. https://doi.org/10.1175/bams-d-11-00094.1

Wang J, Hong Y, Li L, Gourley JJ, Khan SI, Yilmaz KK, Adler RF, Policelli FS, Habib S, Irwn D, Limaye AS, Korme T, Okello L (2011) The coupled routing and excess storage (CREST) distributed hydrological model. Hydrol Sci J 56:84–98. https://doi.org/10.1080/02626667.2010.543087

Xian SY, Lin N, Hatzikyriakou A (2015) Storm surge damage to residential areas: a quantitative analysis for Hurricane Sandy in comparison with FEMA flood map. Nat Hazards 79:1867–1888. https://doi.org/10.1007/s11069-015-1937-x

Xue X, Hong Y, Limaye AS, Gourley JJ, Huffman GJ, Khan SI, Dorji C, Chen S (2013) Statistical and hydrological evaluation of TRMM-based multi-satellite precipitation analysis over the Wangchu Basin of Bhutan: are the latest satellite precipitation products 3B42V7 ready for use in ungauged basins? J Hydrol 499:91–99. https://doi.org/10.1016/j.jhydrol.2013.06.042

Zahmatkesh Z, Karamouz M, Goharian E, Burian SJ (2015) Analysis of the effects of climate change on urban storm water runoff using statistically downscaled precipitation data and a change factor approach. J Hydrol Eng 20:05014022. https://doi.org/0501402210.1061/(Asce)He.1943-5584.0001064

Zhao T, Shao Q (2015) Detecting floodplain inundation based on the upstream–downstream relationship. J Hydrol 530:195–205. https://doi.org/10.1016/j.jhydrol.2015.09.056

Zhu Z, Oberg N, Morales VM, Quijano JC, Landry BJ, Garcia MH (2016) Integrated urban hydrologic and hydraulic modelling in Chicago. Illinois Environ Model Softw 77:63–70. https://doi.org/10.1016/j.envsoft.2015.11.014

Chapter 9
Planning for Community Resilience Under Climate Uncertainty

Ross B. Corotis

Abstract Community resilience requires an accurate estimate of the stressors to which that community could be subjected, and the likelihood of their occurrences and magnitudes. Causation of natural hazards can be categorized conveniently into four general classifications: hydrological, climatological, meteorological and geophysical. For all of these categories, future risk generally has been based on probability models calibrated from past experience. But for the first three, climate uncertainty demands a reexamination of that approach. This reassessment is particularly important for communities subjected to riverine and coastal flooding. In this chapter we address the vulnerability of flood-prone communities as a whole, rather than their individual structures. We look at community vulnerability, first to dams and levees and then to coastal flooding, and then introduce the concept of adaptive management for a changing climate. Finally, we examine the importance of incorporating future cots in community decision-making.

Introduction

Future climate change is expected to cause rising sea levels and changes in the number and characteristics of strong storms, especially tropical hurricanes (Wuebbles 2016, Rosowsky et al. 2016). This means that coastal communities in particular may be at very different risk levels than assumed in the past (Ellingwood and Lee 2016; MacLean 2016). In addition, communities developed with the assumption of safety from dams and levees may see very different demands on these structures than those that were expected when they were constructed. This chapter focuses on communities as a whole, rather than individual structures or infrastructure, outlining participatory methods by which community mitigation and adaptation actions can be planned. These steps are crucial as communities develop

R. B. Corotis (✉)
Department of Civil, Environmental and Architectural Engineering,
University of Colorado, Boulder, CO 80309-0428, USA
e-mail: corotis@colorado.edu

strategies to become more resilient, particularly in an environment of climate variability. Since the high degree of uncertainty makes it difficult to predict precisely which stakeholders are likely to be impacted, traditional approaches involving only risk professionals are not suitable, and instead a broad array of affected stakeholders should be involved (Stewart 2016a). In addition, this climate variability means that traditional approaches of calibrating risk probabilities from historical data may no longer be adequate.

Community Vulnerability and Resilience Principles Related to Dams and Levees

Many communities in the world are vulnerable to flooding due to their location downstream of dams or cross-stream of levees. Flooding can be caused by failure of these structures to perform as designed (a likelihood often increasing over time due to inadequate or improper maintenance and repair), or by higher than anticipated hydraulic demands. It is particularly in this latter category where changing climate can result in dramatic changes in the likelihood of occurrence of major events. For instance, facilities are typically designed to resist a hydrologic event for which there is a computed probability of occurrence for an event of that size or larger annually. This is referred to as the annual exceedance probability, and climate uncertainty could lead to events far in excess of those planned or assumed. For instance, extreme coastal storms (those costing more than a billion dollars of damage, adjusted to 2013 dollars) increased from 0.4 per year (i.e., on average one every two-and-a-half years) in the early 1980s to more than one a year in the period 2000–2013 (National Research Council 2014).

In many cases, these communities are uninformed of their vulnerability. The public at large may even be unaware of the existence of these structures, and the fact that they bring ongoing benefits to the community (such as their disaster-preventing function) as well as risks. In risk and resilience studies it is common to define risk as the product of the probability of occurrence of an event (referenced to some time period of exposure), multiplied by a measure of the consequences of that event (typically in monetary terms, although it often may be preferable to separately account for monetary losses, morbidity, mortality and duration of impact). Even an informed community likely lacks a full understanding of the risks of failure to the physical, economic and social infrastructure, and the fact that these might extend far beyond the zone of potential flooding. In addition, members of the community may not comprehend fully the wide-ranging benefits and risks of these structures; issues that must be part of a cost-benefit analysis requisite to determining and justifying construction, repair, maintenance and upgrading expenditures.

Exacerbated by the changing risk levels due to climate change, it is important that communities prepare for and mitigate against the consequences caused by the potential failure of these structures to protect the assets of each affected community. Dam and levee safety professionals are but one set of players, but a critical one, in the overall picture of community resilience. The governance of dams in the United States is guided by the federal National Dam Safety Program, which regulates individual dam owners and programs; there is no corresponding program for levees, although for federal dams the U.S. Army Corps of Engineers provides guidance. This section focuses on the processes and principles that will lead to increased community resilience, and recommendations in this section are in line with a recent report by the U.S. National Academies, of which the author of this chapter was a contributing author (National Research Council 2012).

It is important to note that while the issues considered below are common to communities behind either dams or levees, these two types of structure differ in many aspects. For instance, in the United States a National Dam Safety Program, along with individual state protocols, forms the basis for regulation of dams, both public and private. No such coordinated effort exists for levees except for those under federal responsibility.

As mentioned above, a community program to assess and improve resilience must encompass the dam and levee professionals (including owners, operators and regulators) as well as all other stakeholders, both individual and institutional, that are affected by the presence of the structures, and would be impacted by its failure. These should include persons and property owners as well as organizations potentially experiencing direct consequences (such as loss of property or life due to inundation), as well as those with indirect consequences such as financial loss, service interruption and ecological damage. It is important to define these groups broadly, including, for instance, commercial and manufacturing enterprises, banks, real estate developers, insurance companies, government organizations, social-ecological networks, cultural resources and members of the wider economy (National Research Council 2012). Particularly since climate change introduces a high level of uncertainty, it is important for both the professionals and the community members at large to work together to understand the range of potential outcomes, expressed both probabilistically and in terms of degree of possible damage. It is also important for the community at large to understand that damages are a function not just of the dam or levee, but of community decisions such as land use planning, social constructs and building code regulations. Examples could include preservation of flood-prone areas for natural reserves or recreational and sports complexes (taking into account warning and evacuation times), requirements for larger drainage structures, and increase in elevation of base occupancy levels for buildings. Appropriate accounting of future risks (a recommendation will be made later in this chapter) needs to be understood by all, even when it is not possible to accurately quantify those risks due to uncertain climate change scenarios.

The desire to involve a wide range of community participants in risk-based decisions implies that these individuals need access to all available information and analyses. The lack of exactitude among professionals regarding climate changes means that it may no longer be possible to just take technical information as the basis for decision. Instead, the uncertainty inherent in these changes must be shared with the members of the public. Qualitative issues of intra- and intergenerational social equity, social vulnerability and environmental preservation may be important aspects of such decisions. At the same time, concern over the past decade or so with terrorist intentional harm has caused the U.S. government to restrict access by the public to certain infrastructure information, such as pipeline locations and inundation potential. These conflicting concerns must be balanced if community players are to be full partners in community resilience planning. Considerations include the increased risk of terrorist disruption to physical infrastructure systems (transportation, energy, communication, water, wastewater treatment, etc.) when their vulnerabilities are widely known, versus the advantage of increased willingness of the community to invest resources addressing such vulnerabilities following widespread awareness. Goals of national security, proprietary interests and liability concerns must be balanced with the goals of a community to enhance risk assessment, preparedness, mitigation, response, recovery and the capacity for resilience (for more discussion see Cox and Cox 2016; Gilbert et al. 2016; Hansson 2016).

Climate change issues represent a unique situation with respect to community risk and infrastructure management. The widespread degree of long-term uncertainty is unprecedented across such a broad range of potential impacts. Analogies can be drawn, for instance, in designing long-span suspension bridges when aeroelasticity was not understood, or developing watershed pacts when there was uncertainty of annual flows. In the former case, the Tacoma Narrows Bridge, constructed in 1940, crossed Puget Sound in the state of Washington. It was the third longest suspension bridge in the world (after the Golden Gate bridge in San Francisco and the George Washington bridge across the Hudson river between New York City and New Jersey), and represented the slenderest profile (cross section). As such, it extended the art of suspension bridge design beyond then-current experience. Static studies (non-dynamic) indicated that the bridge would be stable, but the dynamic effect of resonance was not well understood at the time, and it was only after the collapse (shortly after construction) that the nascent field of bluff body aeroelasticity was recognized for its importance (Simiu and Scanlan 1986). In the latter case, the Colorado River Pact was drawn up in 1922 to apportion the flow of the Colorado River to the states of Arizona, California, Colorado, Nevada, New Mexico, Utah and Wyoming. The allocations were based on historical records available at the time, dating from 1905 to 1922. Subsequent records have indicated that this period was one of unusually high precipitation, and the actual annual average might be as much as 20% lower (Wikipedia 2016). In contrast to climate change uncertainties, however, in these cases the lack of knowledge was focused, and the consequences limited in extent. Sensitivities to climate change unknowns require both facility managers and community decision-makers to be aware of the impacts in each other's milieu. This collaborative cultural awareness of shared impacts will be necessary if a truly collective risk-management process is to be developed for community sustainability and resilience with respect to flooding behind structural facilities. There are many existing tools that can assist in this effort, and one of the most effective is the Maturity Matrix for Community

Engagement (National Research Council 2012). The idea for benchmarking and managing collaboration and engagement among community representatives can be developed from the Capability Maturity Model Integration concept created for the software and systems engineering industry (Paulk et al. 1994), and implemented for resilience in communities behind dams and levees (Bennett and Sykes 2010). It offers a way to plan and gauge community engagement in major decisions where uncertainty plays a dominant role. The Maturity Matrix for Assessing Community Engagement (National Research Council 2012) has multiple levels of increasing engagement for various elements of community risk and resilience. For instance, consider the element of Emergency Action Plans, which consists of tools for emergency planning for preparedness, response and recovery. Activity Level I might indicate no activity, Level II could be comprised of emergency action plans developed by the dam or levee owner, Level III would be the development of these plans with input from an emergency management agency, Level IV would incorporate input from community members and other stakeholders, and finally Level V would involve community collaboration to develop a program that reflects inherent community values. One aspect that adds to the complexity of the situation is that while there is a need for public engagement, the public often does not have sufficient knowledge/understanding of probabilities etc. This also bring in the issue of communicating risk in a clear and unbiased way (McCarthy and Sugrue 2016; Woo 2016).

Community Vulnerability Due to Coastal Flooding

Communities that are not protected by dams and levees include coastal ones that are susceptible to flooding. Climate projections suggest that these communities may be especially vulnerable to hurricanes and other coastal storms, and of course in the case of Small Island Developing States (SIDS) the entire country might be at heightened risk. Current development plans and structural protection strategies have generally been established under the assumption of stationary demand, whereas climate projections suggest possible increases in variability and frequency of such major storms affecting coastal communities. Defenses include structures such as sea walls and storm surge barriers and nature-based protection such as dunes and beach nourishment. Other measures call for decreasing the vulnerability through land use planning, reducing the number of people at risk or designing facilities to be able to withstand flooding. This section follows recommendations of a recent report by the U.S. National Academies, of which the author of this current chapter was a contributing author (National Research Council 2014).

A comprehensive assessment of the effectiveness of past coastal protection strategies needs to be performed, and then the sensitivities of these methods to changing sea levels and storm surges consistent with climate change models should be completed. Such an analysis must include the main categories of community protection:

- Nature-inspired methods such as beach nourishment and dune-building
- Structures such as sea walls and storm surge barriers
- Ecosystem features such as mangroves, coral reefs and barrier islands
- Land use zoning such as setbacks and coastal zone restrictions
- Building modifications such as raised elevations.

The first of these is premised on the concept that natural landforms provide a fundamental defense to coastal waves, storm surge and flooding. These, however, require sufficient space, and are inherently evolving. Therefore, augmentation of natural landforms may be necessary to provide stability. Common methods include beach nourishment and dune building, and success has been seen with the combined efforts of beach nourishment to increase width and stability from erosion, along with increased width and height of dunes. These actions have also increased tourism by making beaches more attractive, but have decreased the value of properties that have lost views of the ocean or are more removed from the water's edge. They require consideration of the changed bathymetry in areas from which the sand is dredged, and have also led to increased sand deposition by natural wave action, altering neighboring areas. These latter concerns should be addressed by a regional view, rather than a single project basis.

Structural approaches have a long history of successful implementation, especially in concert with nature-based methods, but there are few data on their cost-effectiveness. They also have impacts of changing the beach profile, increasing erosion in unprotected areas, and interfering with the natural flow of sediment. They may also have an effect on the coastal marine biota. An alternative to sea walls is the use of coastal levees, which are located landward and designed either to dissipate wave energy or to minimize flooding caused by costal storm surge along rivers. These also have impacts due to their modification of natural water exchange for lands along the river. Finally, storm surge barriers are not in common use, but the Thames barrier in the United Kingdom is an excellent example of such a project.

Ecosystem approaches have received a relatively recent surge in interest, but the extent of space necessary and the requirement for strict zoning regulations to ensure their effectiveness means that they should be considered part of a broad, regional approach. Their success is very much dependent on the width available perpendicular to the shoreline, and they are more effective at diminishing wave action than storm surge.

Land use planning has shown high benefit to cost ratios, but is generally considered to fall under the authority of the local jurisdiction, where the pressures of development may be strongest. Tax revenue and tourism are typically at short-term odds with hazard zoning, land purchase and setbacks.

Finally, building elevation regulations and building structural strengthening can be effective, but involve careful consideration of private property rights and the takings clause of the U.S. constitution.

Separate from these five strategies, public information campaigns and guidelines for elevation of new structures tend to avoid pushback from sensitive political and economic constituencies.

Coastal communities often lack a central agency responsible for assessing and managing flooding risk. Guidelines that have emerged over the decades may have seemed adequate under steady state conditions, but are incapable of dealing with the multiple challenges associated with climate change. In particular, elected officials and developers benefit from the immediate rewards of development, but are not held responsible for the long-term costs to the community. While such a challenge is not unique to climate change, it is of particular concern in that case since residents of a community may not have the information to understand the potential changes in exposure and risk that are commensurate with changing climate; events that they have never experienced. In the United States, federal disaster relief funding has created a moral hazard, implicitly encouraging communities to build in vulnerable areas, and then providing financial relief when a disaster occurs. Funding for pre-disaster planning has generally been meager in relation to post-disaster relief, and certainly not focused on the increased annual likelihood of flooding due to potential climate change effects. In the U.S., coastal risk has traditionally been assessed by the U.S. Army Corps of Engineers (USACE), but this agency has no funding or mandate to assess coastal risk at a regional or national scale.

Federal programs in the United States are often justified on a benefit-cost basis, and indeed this is required for U.S. Army Corps of Engineers projects. While such an approach can be valid in many cases, it is sometimes challenging to assess comprehensive costs and benefits across the spatial and temporal scales associated with community resilience and sustainability. It is also often very difficult in a comparative analysis to assign monetary cost to such issues as mortality and morbidity, and to assign monetary benefits for improved quality of life, long-term health benefits, ecological preservation, historical value and personal satisfaction. There are valid arguments that this might not be the most appropriate tool for projects that involve important social impacts (OECD 2007; Rowell 2016). It may also be that the optimal choices from a benefit-cost analysis result in risks to individuals and communities that are not considered to be in an acceptable range. An alternative approach is to use a risk-consistent analysis to develop policies that inform various communities of the relative risks, and then provide a collection of strategies to meet an acceptable, consistent risk level. In a pure benefit-cost analysis, one would be expected to compare alternative strategies for flooding protection for instance, and then select the one that has the highest benefit-cost ratio. In a risk-consistent approach, one would investigate various levels of these flood protection strategies, in each case selecting a level that gives a constant level of risk (recalling this is the probability of occurrence multiplied by a measure of the extent of consequence). One would then select the least cost alternative among these risk-consistent options. In summary, the selection of a criterion upon which alternative strategies may be compared is a subject of significant investigation and debate (Gardoni and Murphy 2014).

Many of the policies developed in the United States are based on assuming a 100-year return period level of safety. This hazard-consistent approach does not meet the guidelines of either of the approaches in the preceding paragraph, and is also misunderstood by many people who do not appreciate the stochastic nature of the recurrence interval. The basic concept is to design facilities to be able to withstand an event for which there is an annual occurrence probability of 0.01 for an event of at least that magnitude. Such an approach ignores the cost of such design, and the tradeoffs of risk and cost for alternative approaches and levels of safety. Such an event is often referred to as the 100-year event because in the case of a stationary sequence with independent occurrences, such an event has an expected return period of 100 years (i.e., on average it can be expected to occur once every 100 years). Such an assumption of independent occurrences, however, implies that the chance of experiencing such an event is independent of the number of years that have passed since the last occurrence. These assumptions for an independent stationary series are likely to be violated in the case of climate change.

The issues and approaches discussed in this section are important, but they have generally lacked a regional or national vision for their execution. For instance, beach nourishment has not included sediment transport models that incorporate management strategies for large regions. They also have not provided a strong reward system for current community decision-makers to implement policy requiring current cost requirements and only future potential benefit savings. This last issue is especially important when the uncertain potential changes due to climate variability and trends are included. Not only might the 100-year return period event based on historical records become one with a much shorter recurrence period, but the level of epistemic uncertainty associated with these events could be greater than has been assumed in the past.

In summary, both cost-benefit and risk-consistent approaches are based on fundamental concepts whose implications are traditionally and meaningfully interpreted in terms of stationary sequences calibrated to past, experienced events. The current state of knowledge with respect to climate change strongly supports the conclusion that stationarity is not valid, and that the degree of uncertainty for future trends is high. The former can be incorporated into traditional approaches with adjustments for nonstationary, but the concept of return period or annual risk then becomes time-dependent also. Moreover, the high degree of uncertainty, the expected nonlinearity of community resilience measures to climate drivers, the significant level of consequence to society, and the practical irreversibility of many decisions over infrastructure lifetimes means that concepts of the precautionary method may be appropriate. Such a method has not been common with the U.S., but it is intended precisely for such situations. The United Nations states, "In order to protect the environment, the precautionary approach shall be widely applied by States according to their capabilities. Where there are threats of serious or irreversible damage, lack of full scientific certainty shall not be used as a reason for postponing cost-effective measures to prevent environmental degradation." (United Nations 1992). The fundamental precept of the precautionary principle is the concept that if a risk management decision may cause negative consequences, and there is an absence of scientific consensus, the onus is on demonstrating that the action does not cause this risk. Other very different approaches from what has been used in the past may be needed. One of those, based on the concepts of Generalized Information Theory, introduces uncertainty measures that are much more robust than those of probability theory, such as monotone measures of belief, plausibility and basic evidence assignment (Corotis 2016).

Adaptive Management for a Changing Climate

Due to recent federal legislation in the United States, the Federal Emergency Management Agency (FEMA) has been re-examining the risk basis for actuarially-based flood insurance rates. By standard accounting practice, insurance rates should be based on the actual computed risk (adjusted as experience on payouts is accumulated), providing margins for reserves and operating expenses. When coverage is divided into different classes, these classes should share a high degree of similarity of risk, and the differences in premiums between classes should realistically reflect the variances in expected payouts. The National Flood Insurance Program (NFIP) was created in 1968, and was intended to both provide flood insurance coverage to individuals and to promote more flood-resilient development (AIR 2005). Rates for new structures were intended to be actuarially-based, reflected the risk of flooding. Rates for existing structures, however, were subsidized at below-actuarial rates, so as not to place an immediate burden on current property values. Currently, about one-fifth of the policies written by NFIP receive below-actuarial subsidized rates (National Research Council 2015). The Biggert-Waters Flood Insurance Reform Act of 2012, and then the Homeowner Flood Insurance Affordability Act of 2014 both called for a sound theoretical basis for flood risk calculations, and the eventual financial self-sufficiency of the National Flood Insurance Program. The second Act, delayed implementation of actuarially-based rates by several years, and limited the annual increase in rates, thereby diminishing the short-term impact on individuals. In the long-term the eventual goal of both Acts is the same, however.

In addition to making affordable flood insurance available to homeowners and businesses, NFIP was also intended to reduce their reliance on federal disaster relief funds. A dual goal was to assist people in becoming informed of their flood risk, and thereby to encourage development and building practices that reduce community vulnerability. The grandfathering of existing structures at insurance rates below risk-based market values, and the subsidization of rates when updated flood maps show an increased risk for a property, have had an unfortunate deleterious effect on this second goal. Thus increased construction in flood-prone areas has led to an increase in property damage in the United States due to flooding over the past half century. In the 1950s there were 5–10 flood-related Presidential Disaster Declarations a year, and since the year 2000 these have varied between 20 and 50 (http://www.fema.gov/disasters/grid/year) (accessed August 5, 2015). This section will focus on recommendations of a recent report by the U.S. national Academies, of which the author of this current chapter was a contributing author (National Research Council 2015).

Currently, flood risk in the United States is tied to the 100-year event, with extrapolations made to the 20-year and 500-year events. These extrapolations depend on the watershed and floodplain characteristics where structures are located, and whereas they are actually unique to each watershed, approximations of grouping have been made for actuarial simplification. While any hydraulic approach could in theory be adjusted for climate change, the method currently used by FEMA offers particular challenges because it is so heavily tied to a single event, the 100-year flood. As discussed earlier, this concept of return period is likely to lose much of its physical interpretation in a time of climate change. Further, climate variability is likely to affect the relative levels of flooding for different return periods. For instance, say a property has a topography such that the current flood elevation associated with a 200-year return period (an annual probability of 0.005) is one meter higher than that of the reference 100-year flood. Under a revised climate scenario, the 200-year event may be two meters higher than the reference 100-year flood. Thus the relative risk for different elevations can be expected to change. In order to give the reader a general idea of the intricacies of such calculations, the next paragraphs will explain the NFIP approach in basic detail.

Flood risk is based on separation of the probability of flooding above a particular elevation, and the expected damage for a given flooding level. The first calculation is based on what are termed the PELV curves (PELV is shorthand notation for probability of water surface elevation exceedance). These curves were developed in the 1970s, and are anchored at the 100-year flood level, and the shape of the curve depends primarily on the difference between the 100-year and 10-year flood depths (the 1% and 10% annual chance of exceedance). Calculated flooding risk is especially sensitive to the shape of this curve for structures with lower elevations (those whose elevations place them in the category of having an annual flood risk exceeding 1%, and for which insurance payouts for damage are often dominated by structures with greater than a 10% annual chance of flooding). Changing climate conditions are likely to subject a lot more structures to this more frequent flooding, moving structures from risk categories on the order of 1% annual probability to values closer to 10%. And many structures already experiencing more frequent flooding than that associated with the 100-year return period events may see more frequent flooding. For these structures, the precise characteristics of the floodplain will be even more important, and greater granularity than currently used by NFIP will be called for. While it is true that there also likely will be structures moved into lower risk categories (such as from 1% to 0.2%), these structures do not dominate the payout portfolios. Therefore, the total experienced losses are likely to increase.

The second calculation is based on curves relating the damage (as a percentage of the replacement value of a structure) to the depth of flooding. These are called the DELV curves, shorthand notation for depth of flood elevation. They have been derived from hydrologic studies by the U.S. Army Corps of Engineers, and updated by actual claims data using the concept of Bayesian updating. The DELV curves do not reflect such factors as duration of flooding, water contamination or speed of moving water, and therefore it is not clear how changes in climate conditions for sea level rise and storm magnitude will affect these curves. In addition, with more structures likely to experience flooding due to climate changes, a more refined model of building characteristics and their relation to damage would be advisable (e.g., type of interior and exterior finishes, usage of basements, foundation properties and local drainage measures). An excellent example is a home known to the author built within the 100-year floodplain in Boulder, Colorado. There is an "upstream" door and a "downstream" door, connected by a stone hallway that is at a lower elevation than the rest of the first floor (there is no basement). In the case of a flood, the doors are opened, sandbags are provided to increase the elevation jump from the hallway to the rest of the living space, and the water flows through with minimal damage.

As with any actuarial enterprise, the NFIP uses historical claims records to update its risk-based rates. With climate change, it will be necessary to move from the traditional approach of basing calibration of future probability distributions purely on historical data. Not only might averages and variances be dissimilar in the future, but the actual form of the distributions might be different. Nonstationary approaches such as moving average autoregressive (ARMA) models can account for changes in the average and standard deviation, as well as possibly higher moments (especially skewness), but they have not generally been proven for highly nonlinear processes that might affect the entire distribution.

An interesting question is whether calculated future losses with climate change can be accommodated by adjustments to the current flood damage approach used by the NFIP, or whether a more holistic total risk-based approach is necessary. For instance, the current assumption is that levees that do not meet the certification requirement for a 100-year event are neglected completely, whereas those that do meet certification are assumed to contain no risk of failing for events less than the 100-year ones. Neither is completely correct, in that an uncertified levee will still provide protection for less severe events, and a certified one still contains some probability of not performing completely satisfactorily in events for magnitudes up to the 100-year ones. With the current approach, certification may change due to predicted flooding events, leading to a large jump in computed risk, whereas in a total risk approach the change would be incremental. In addition to levees, similar considerations hold for such facilities as reservoirs, floodwalls and diversion channels. A comprehensive risk-based analysis would take into account all aspects of the rainfall modeling, the watershed and floodplain characteristics, flood protection structures, and the properties of the individual structures. The benefit-cost and risk-consistent principles mentioned earlier are helpful decision criteria upon which to select among alternative options, whereas, "A comprehensive risk assessment would describe risk over the entire range of flood hazard conditions and flood events, including the large, infrequent floods that cause substantial losses to the NFIP portfolio, and the smaller, frequent floods that make up a significant portion of loss…" (National Research Council 2015). Such an approach takes into account all uncertainties, and may be the best way to incorporate various climate scenarios and their effect on precipitation levels (both spatially and temporally).

Another concern is a consequence of the fact that the most accurate estimates may indicate significantly greater flooding damage with better climate-based modeling. This could lead to considerably higher premiums on the part of insurance companies, resulting in short-term decisions by homeowners and business not to purchase flood insurance. Such actions would intensify the level of uninsured losses, and it could be many years before personal flood purchase decisions reflect the true risks (Molk 2016).

A previous exploratory study investigated the concept of adapting infrastructure design for changes in climate (Rajagopalan et al. 2004). The research was not oriented toward long-term climate change, but can serve as an example of how to adjust for climate change A stochastic nonparametric framework permitted annual wind data to be used to estimate a probability density function, which was then convolved with structure fragility curves so that results could be compared in relation to the El Nino Southern Oscillation (ENSO) in the Pacific Ocean. Significant changes in the structural reliability of various types of wood frame structures were observed for the coastal region of North Carolina for the El Nino versus La Nina years. Another study addressing sustainable and resilient infrastructure systems (Lounis and McAllister 2016) makes clear that consideration of changing conditions over the system lifecycle is necessary for realistic performance prediction. A study on the effect of extreme winds on metal structures demonstrates how models can be adjusted to account for climate change scenarios (Stewart 2016b).

Incorporating Future Costs in Community Decision-Making

Effective community-based decision-making requires a way for the long-term rewards and obligations of today's decisions to be conveyed in an open and transparent manner. Only by some such mechanism will the rewards for immediate decisions support the best decisions for sustainable communities. This is necessitated because elected and appointed community decision-makers typically serve for less than a decade, whereas infrastructure lifetimes are 100 years or more, and climate uncertainty is great over this longer lifetime. It is well known that the pressures on community decision-makers are for actions that show immediate demonstrable benefits, leading to re-election or reappointment (Bonstrom et al. 2012).

In considering lifecycle benefits and costs for civil infrastructure, it is imperative that future values be properly discounted. While such a procedure is relatively straightforward for financial quantities, and comparatively insensitive for lifetimes of a few years, neither of these conditions dominate for civil infrastructure. Consequences of infrastructure performance (and failure) include morbidity and mortality, historical significance, environmental protection, biodiversity, and social equity and justice (both intra-generational and inter-generational). Arguments can be made for not discounting some of these, or at least for using a very low rate, such as for mortality (Pandey and Nathwani 2003), or even a negative rate, such as for historic structures. When considering the long lifetimes of infrastructure, lifecycle benefits and costs are extremely sensitive to the chosen discount rate, and the effects may become especially critical for long-term climate change scenarios (Corotis and Gransberg 2006; Cox and Cox 2016; Farber 2016).

This reality requires a protocol that accounts for the future probability-weighted costs of current infrastructure, and how these would change with investment (Corotis 2002). This is the logic behind CRISP, a proposed Community Resilient Infrastructure Sustainability Protocol. The CRISP report card would be created at election time, and would provide a net present value analysis of the public infrastructure, including not only the current benefit/cost analysis for any new structures, but also any change in value to existing infrastructure. Such an analysis should be stochastic in nature, and reflect current and future benefits as well as costs, including maintenance, repair, environmental degradation, and damage consequences (mortality, morbidity and direct and indirect economic) due to normal and disaster scenarios. The credits and debits (benefits and costs) of existing and planned infrastructure, properly discounted and weighted by likelihood of occurrence over a reasonable lifetime horizon, are the essential ingredients of CRISP. It would provide a linkage between immediate decisions and future returns.

One of the issues with infrastructure lifetime and climate scenarios is the selection of a design lifetime. Current design codes often use a 50-year design life for buildings, 75 years for bridges, and 100 years for lifeline infrastructure. But much of the infrastructure is not removed at the end of its design life; rather it is upgraded and repaired. If one takes the common discount rate of 4%, then the present value for 75 years is 95% of the infinite lifetime value, and after 100 years it is equal to 98% of the infinite value. Therefore, the question of design lifetime can be avoided by using an infinite lifetime, while introducing an error of only a few percent (Vacheyroux and Corotis 2013). This allows discussion to proceed without having to choose an arbitrary time horizon, and removes debate on lifetime from the calculation. This approach could be particularly useful when incorporating climate change scenarios. Rather than focusing discussion on a single future year, with all the concomitant arguments associated with why that particular year was selected, all future years can be automatically included. The increasing uncertainty with more distant years is ameliorated by the discounting factor.

Conclusions

Current risk-based estimates for flooding damage are based on individual structures, and have been created from a mixture of existing methods. With increased flooding likelihood due to climate change, and given increased uncertainty in predictions, communities will likely face even greater challenges in dealing with the calculations of sustainability and resilience. Several of those issues are enumerated in this chapter, and some suggested actions described. Much more research, as well as methods of collaborative deliberation, will be necessary to improve the ability of communities to deal with flooding risks under a changing climate.

References

AIR (2005) A chronology of major events affecting the national flood insurance program. American Institute for Research, 86 pp. Available at https://www.fema.gov/media-library/assets/documents/9612?id=2601

Bennett T, Sykes C (2010) Improving communications within a dam safety program using a maturity matrix approach. Canadian Dan Association Annual Conference, October 2–7, Niagara Falls, Canada. Available at www.cda.ca/proceedings%20datafiles/2010/2010-6a-04.pdf. Accessed for members only 4 Aug 2015

Bonstrom H, Corotis R, Porter K (2012) Overcoming public and political challenges for natural hazard risk investment decisions. J Integr Disaster Risk Manage 2(1):1–23

Corotis RB (2002) The political realities of life cycle costing. In: Casas JR, Frangopol DM, Nowak AS (eds) Bridge maintenance, safety and management, International center for numerical methods in engineering, Barcelona, Spain, CD Rom

Corotis R (2016) An overview of uncertainty concepts related to mechanical and civil engineering. In: ASCE-ASME journal of risk and uncertainty in engineering systems, Part b mechanical engineering, December, 1(4), 040801-1–040801-12

Corotis R, Gransberg D (2006) Adding social discount rate to the life-cycle cost decision-making algorithm. J Reliab Struct Mater 2(1):13–24

Cox LA Jr, Cox ED (2016) Intergenerational justice in protective and resilience investments with uncertain future preferences and resources. In: Gardoni P, Murphy C, Rowell A (eds) Societal risk management of natural hazards. Springer, Berlin, Germany

Ellingwood BR, Lee JY (2016) Managing risks to civil infrastructure due to natural hazards: communicating long-term risks due to climate change. In: Gardoni P, Murphy C, Rowell A (eds) Societal risk management of natural hazards. Springer, Berlin, Germany

Farber DA (2016) Discount rates and infrastructure safety: implications of the new economic learning. In: Gardoni P, Murphy C, Rowell A (eds) Societal risk management of natural hazards. Springer, Berlin, Germany

Gardoni P, Murphy C (2014) A scale of risk. Risk Anal 34(7):1208–1227

Gilbert R, Habibi M, Nadim F (2016) Accounting for unknown unknowns in managing multi-hazard risks. In: Gardoni P, Murphy C, Rowell A (eds) Societal risk management of natural hazards. Springer, Berlin, Germany

Hansson SO (2016) Managing risks of the unknown. In: Gardoni P, Murphy C, Rowell A (eds) Societal risk management of natural hazards. Springer, Berlin, Germany

Lounis Z, McAllister T (2016) Risk-based decision making for sustainable and resilient infrastructure systems. J Struct Eng 142(9). September, 1943-541X.0001545, ASCE

MacLean D (2016) Climate change and natural hazards. In: Gardoni P, Murphy C, Rowell A (eds) Societal risk management of natural hazards. Springer, Berlin, Germany

McCarthy TG, Sugrue NM (2016) Social choice and the risks of intervention. In: Gardoni P, Murphy C, Rowell A (eds) Societal risk management of natural hazards. Springer, Berlin, Germany

Molk P (2016) Private versus public insurance for natural hazards: individual behavior's role in loss mitigation. In: Gardoni P, Murphy C, Rowell A (eds) Societal risk management of natural hazards. Springer, Berlin, Germany

National Research Council (2012) Dam and levee safety and community resilience. The National Academies Press, Washington, D.C.

National Research Council (2014) Reducing coastal risk on the East and Gulf Coasts. The National Academies Press, Washington, D.C.

National Research Council (2015) Tying flood insurance to flood risk for low-lying structures in the floodplains. National Academies Press, Washington, D.C.

OECD–International Transport Forum (2007) Joint transport research centre discussion paper Nol 2008–6. Round Table, 25–26, October 2007, Boston. http://www.internationaltransportforum.org/jtrc/discussionpapers/DP200806.pdf Accessed 4 Aug 2015

Pandey MD, Nathwani JS (2003) Discounting models and the life-quality index for the estimation of societal willingness-to-pay for safety. Reliability and Optimization of Structural Systems, Balkema, London

Paulk PC, Weber CV, Curtis B, Chrissis MB (1994) The capability maturity model: guidelines for improving the software process. Carnegie Mellon University Software Engineering Institute, Addison-Wiley Professional, Boston, MA

Rajagopalan B, Edward O, Corotis RB, Frangopol DM (2004) Estimating structural reliability under hurricane wind hazard: applications to wood structures. In: Proceedings of the ninth ASCE joint specialty conference on probabilistic mechanics and structural Reliability, July 26–28, Albuquerque, NM CD-ROM

Rosowsky DV, Mudd L, Letchford C (2016) Assessing climate change impact on the joint wind-rain hurricane hazard for the Northeastern U.S. Coastline. In: Gardoni P, Murphy C, Rowell A (eds) Societal risk management of natural hazards. Springer

Rowell A (2016) Theories of risk management and multiple hazards: thoughts for engineers from regulatory policy. In: Gardoni P, Murphy C, Rowell A (eds) Societal risk management of natural hazards. Springer, Berlin, Germany

Simiu E, Scanlan RH (1086) Wind effects on structures. Wiley, New York

Stewart M (2016a) Risk and decision-making for extreme events: climate change and terrorism. In Gardoni P, LaFave J (eds) Multi-hazard approaches to civil infrastructure engineering. Springer, Berlin, Germany

Stewart M (2016b) Climate change impact assessment of metal-clad buildings subject to extreme wind loading in non-cyclonic regions. Sustain Resilient Infrastruct 1(1–2):32–45

United Nations (1992) Report of the United Nations conference on environment and development (Rio de Janeiro, 3–14 June), Annex I, RIO declaration on environment and development, available at www.un.org/documents/ga/conf151/aconf15126-1annex1.htm. Accessed 25 Aug 2016

Vacheyroux G, Corotis R (2013) Strategies of investment in the management of urban bridges: a life-cycle approach illustrated for Paris. Struct Infrastruct Eng Maintenance Manage Life-Cycle Design Perform 9(11):1080–1093

Wikipedia (2016) https://en.wikipedia.org/wiki/Colorado_River_Compact, no date. Accessed 25 Aug, 2016

Woo G (2016) Participatory decision-making on hazard warnings. In: Gardoni P, Murphy C, Rowell A (eds) Societal risk management of natural hazards. Springer, Berlin, Germany

Wuebbles DJ (2016) Setting the stage for risk management: severe weather under a changing climate. In: Gardoni P, Murphy C, Rowell A (eds) Societal risk management of natural hazards. Springer, Berlin, Germany

Part IV
Responding to Climate Change: Mitigation and Adaptation

Chapter 10
Climate Change Governance and Local Democracy: Synergy or Dissonance

Emmanuel O. Nuesiri

Abstract This chapter focuses on governance arrangements in the reducing emissions from deforestation and forest degradation, plus the role of conservation, sustainable management of forests and enhancement of forest carbon stocks in developing countries (REDD+) initiative. The United Nations Collaborative Programme on Reducing Emissions from Deforestation and Forest Degradation (UN-REDD Programme) supports developing countries adopting REDD+, and commits to strengthen local democracy as a safeguard such that REDD+ benefits to local people are not captured by elites. The chapter questions whether the UN-REDD funded Nigeria-REDD program meets this safeguard requirement. Research methods included literature review, semi-structured interviews, focus group meetings and participant observation. The study finds that the design of Nigeria-REDD was not inclusive of democratically elected local government authority. The UN-REDD approved the Nigeria-REDD proposal, trusting that NGOs who were involved in designing Nigeria-REDD, will push for democratic governance. However, NGOs do not have a mandate to democratically respond to the needs of local people. The chapter recommends that UN-REDD should not only engage with NGOs, but also with elected local government authority, if it is to strengthen local democracy as a safeguard against elite capture of REDD+ benefits.

E. O. Nuesiri (✉)
Cline Center for Democracy, University of Illinois Urbana Champaign, 2001 South 1st Street #207, Champaign, IL 61820, USA
e-mail: enuesiri@illinois.edu

Introduction

This chapter examines the governance arrangement in a specific global climate change initiative.[1] This is the reducing emissions from deforestation and forest degradation in developing countries (REDD) plus the role of conservation, sustainable management of forests and enhancement of forest carbon stocks in developing countries (REDD+). This initiative is part of the global response to the climate change crisis (Corbera and Schroeder 2011). However, REDD+ could lead to a loss of livelihood for poor forest dependent people (Accra Caucus 2013; Roe et al. 2013). This is because local people will be denied access to forest areas that would be set aside for reforestation and carbon sequestration as part of REDD+. In some cases local people might be displaced from land set aside for REDD+ (Beymer-Farris and Bassett 2012).

In 2008, the United Nations set up the UN Collaborative Programme on Reducing Emissions from Deforestation and Forest Degradation in Developing Countries (UN-REDD Programme), to provide financial and technical support to developing countries interested in setting up REDD+ programs (UN-REDD 2008). In recognition of the potential negative impact on local people of REDD+, the UN-REDD commits to strengthen local democracy as a social safeguards against the marginalization of local people who would be affected by the implementation of REDD+ (UN-REDD 2008; CIF, FCPF and UN-REDD 2010).

This chapter assesses whether the UN-REDD rhetoric matches its activities. It proceeds by first analyzing the UN-REDD governance mechanisms informed by the theory of political representation (see Pitkin 1967; Manin et al. 1999; Mansbridge 1999; Saward 2006, 2008; Urbinati and Warren 2008; Rehfeld 2011; Montanaro 2012). It then analyzes the US$4 million UN-REDD funded Nigeria-REDD readiness program (see Nuesiri 2016). The analyses shed light on whether UN-REDD rhetoric to strengthen local democracy matches its practice.

Research methods were primarily literature review and semi-structured interviews during field work in the summer of 2012 and 2013. A total of 125 research participants drawn from local communities, Nigeria-REDD, UN-REDD and local NGO staff were interviewed. Research methods also included 3 focus group meetings with personnel of 3 NGOs (1 local, 1 national, 1 international), and participant observation while attending a local Council of Chiefs meeting as an invited resource person for a knowledge sharing session about Nigeria-REDD.

The chapter is organized as follows: Sect. 2 is a review of political representation theory and argues that local democratic authorities are the building blocks to establishing democratic governance in any nation. When there are effective local democratic authorities there is a political space in which the citizenry can express needs and demand accountability from elected representatives. Section 3 examines

---

[1]Parts of an earlier version of this paper were published as Nuesiri (2016). Deepening Local Democracy for a more Just Global Governance Regime. ISS Colloquium Paper No. 30, The Hague, Netherlands: International Institute of Social Studies.

the UN-REDD strategy for strengthening local democracy through a close reading of UN-REDD constitutive documents. Section 4 shows that despite being drawn up with the supervision of the UN-REDD, Nigeria-REDD allowed for the exclusion of elected local government authorities from the participatory consultative processes that accompanied the design of Nigeria-REDD readiness proposal.[2]

Section 5 discusses why the UN-REDD tolerated the exclusion of elected local government authorities from the consultative process that validated Nigeria-REDD, while Sect. 6 concludes the chapter with a summary of its findings and recommendations on what the UN-REDD can do to strengthen local democracy for a more just climate change governance regime. The findings of this chapter might be unique to Nigeria, but the recommendations have universal value, especially after an external review of the UN-REDD noted that REDD+ initiatives are weakening community rights over forests (Frechette et al. 2014).

The argument that this chapter is making for the full inclusion of elected local government authorities in Nigeria-REDD program, is not on the premise that these authorities are efficient in delivering their tasks. Like national governments and international organizations, they are besought with problems of mismanagement and corruption. However, these problems have not stopped the UN-REDD from working with national governments. The decision to work with national governments is not so much based on performance but on function. National governments speak for the national territory, likewise local government authorities speak for the local, and are the foundations democratic governance (UNSG 2009).

Political Representation and Local Democracy

Political representation is giving expression to the interests of groups who are physically absent in the course of decision-making (Pitkin 1967; Mansbridge 1999; Urbinati and Warren 2008; Rehfeld 2011). Pitkin (1967) identifies three types of representation—descriptive, substantive, and symbolic.[3] Descriptive representation is when representatives are appointed by a higher authority to stand for a group because they resemble the group (Pitkin 1967); or to put in another, they are appointed by the higher authority because they are considered to be "typical of the larger class of persons whom they represent" (Mansbridge 1999, p. 629).

---

[2]REDD readiness refers to the preparatory and demonstration projects countries need to carry out before implementing a full national REDD+ programme; the readiness proposal is the document showing what these pilot projects would be all about and their accompanying budget (see http://theredddesk.org/encyclopaedia/readiness-preparation-proposal-r-pp).

[3]Pitkin (1967) actually discussed four types of political representation—formal, descriptive, symbolic and substantive; however, her discussion of formal representation was more of a critique of the limited Weberian understanding of representation as deriving from formal authorization of an agent by the state to represent a constituency to the state or to represent the state to an audience.

Substantive representation is when representatives act for and are accountable to the represented; the represented are also able to evaluate and sanction their representatives (Pitkin 1967). Substantive representation is morally superior to descriptive and symbolic representation for the accountability checks it places on representatives (Pitkin 1967). It is also socially just and thus the preferred mechanism behind representative democracy (Grunebaum 1981; Kateb 1981; Manin et al. 1999; Mill 2004; Fraser 2005; Urbinati and Warren 2008; Rehfeld 2011).

In representative democracy, representatives are chosen through elections; are responsive to the interests of the represented; and are downwardly accountable (Manin et al. 1999; Rehfeld 2006). Representation is undemocratic when electoral choice, responsiveness and downward accountability are absent. Where undemocratic authorities choose to be responsive to the governed, this is "good despotism" (Mill 2004, p. 36), or benevolent and or benign dictatorships (Wintrobe 1998; Manin et al. 1999). Democratic representation is a critical instrument for engaging local people in support of initiatives like REDD+ because it is inclusive and a non-violent mechanism for resolving differences (Dahl 1989; Davenport 2007).

Symbolic political representation is when a person or thing such as the flag, or institution such as a non-governmental organization (NGO) represents peoples (or territories) based on shared beliefs, aspirations, norms and world view (Pitkin 1967). Symbolic representatives are appointed following cultural norms or executive order. They are not elected, so are not statutorily mandated to be responsive and accountable to the represented. These representatives legitimize their status and actions by employing iconic images, objects with moral authority, and emotive rhetoric (Edelman 1985; Wedeen 1998). Symbolic representatives sometimes manipulate their affective ties with the represented by inducing support for decisions, which may not be a substantive response to the interests of the represented (Lombardo and Meier 2014). During the apartheid era in South Africa, some tribal authorities urged support for discriminatory land policies though it was not in the people's interest to do so (Ntsebeza 2005).

When symbolic representatives support a response that is attentive to the needs of the represented, this does not make them substantive representatives as their relationship with the represented is still defined by emotions opened to manipulation and not by accountability relations. When elected representatives' support symbolic decisions (see Bluhdorn 2007; Edelman 1985; Miller 2012), as is the case when crafting environmental legislation (Matten 2003; Newig 2007; Stavins 1998), this does not make them symbolic representatives as they remain formally and legally accountable to the represented.

In assessing political representation, it is important to note the distinction between types of representation. Descriptive and symbolic representatives even when they act in the interests of the represented, remain undemocratic because they cannot be held accountable and sanctioned by those they represent. On the other hand, when substantive representatives' respond symbolically to demands of their constituents, they remain democratic representatives because they are accountable to, and opened to sanctions from their constituents.

Edelman (1985) asserts that symbolic politics is used by governments to manipulate the public. Brysk (1995) adds that symbolic politics is also employed by non-governmental organizations (NGOs) to influence corporations, governments and the public (see also Keck and Sikkink 1999; Miller 2012; Silveira 2004). Matten (2003) asserts that symbolic politics is the response of policy makers when designing environmental regulations. Symbolic politics is used by Matten (2003) to refer to situations where policy makers talk tough but fail to take action, or craft policies that do not become law, or roll out strong regulations with weak enforcement, or enact legislation that legitimize practices already adopted by industry. Matten (2003) explains that this is because policy makers wish to be seen to be responsive to citizens' concerns but do not wish to antagonize powerful groups like the corporate sector.

In this regard, Stavins (1998, p. 73) notes how the 'Clinton administration announced with much fanfare in June 1997 that it would tighten regulations of particulates and ambient ozone, but the new requirements do not take effect for eight years', and consequently goes on to argue that such symbolic environmental regulations work because 'voters have limited information, and so respond to gestures, while remaining relatively unaware of details.' Newig (2007) observes that when citizens demand environmental legislation but express an unwillingness to pay for substantive action, policy makers interpret this as a signal for symbolic legislation. Newig (2007) refers to this situation as societal self-deception.

Bluhdorn (2007) shows that governments use symbolic politics for communicating with citizens, for avoiding complexity related to substantive implementation of policy, and also as replacement action (which may or may not be deceitful) to avoid substantive policy response. Cass (2012) shows that political leaders also enact symbolic environmental legislation as an instrument of foreign policy. They desire to be viewed as good global citizens while avoiding the cost of substantive action. Baker (2007) maintains that transnational bodies like the European Union (EU) subscribe to symbolic environmental politics. She asserts that the EU's commitment to sustainable development (a transformational paradigm) is symbolic, because EU operational strategies in dealing with environmental problems are informed by ecological modernization. This is problematic because ecological modernization maintains that greater industrial efficiency will reduce the ecological footprint of business, consequently we will do more with less, and this allows for economic processes to continue as usual (see Hajer 1996; Salleh 2010).

Likewise, in the United States there is significant political opposition to federal and state initiatives responding to climate change, because the business backed 'climate denier' lobby insist that economic processes are not contributing to global warming (Hogan and Littlemore 2009). Thus the US government, unlike its European counterparts, is not able to set mandatory greenhouse gas emissions reduction targets for industry. Instead business is encouraged to set voluntary emissions reduction targets (Larsen et al. 2016). This is soft and symbolic regulation setting, as business cannot be formally held to account by government and society.

This brings us to the question of how accountable is the UN-REDD to local people who will be negatively affected by REDD+? The UN-REDD is an environmental regulation setting regime, for which local rural people in developing countries have limited access to information on their decisions and activities. The UN-REDD maintains that it would seek the free prior informed consent (FPIC) of local communities before implementing REDD+ (UN-REDD 2013a). Stavins (1998), Matten (2003), Baker (2007), and Cass (2012) observed that policy makers opt for symbolic over substantive environmental regulations so as not to antagonize powerful interest groups. Has the UN-REDD followed this pattern in Nigeria? The UN-REDD maintains that strengthening local democracy would prevent elite capture of REDD+ benefits (UN-REDD 2008). Is the UN-REDD strengthening local democracy in Nigeria?

Local level democratic institutions like the local government authority are important because they are the building blocks for democratic governance in a nation. They are also the space where citizens become proficient in voicing needs and demanding accountability from elected leaders (Sisk 2001; Coleman 2005). Formal accountability mechanisms for holding local elected leaders to account may not always be effective, but they are statutory, so elected leaders must pay attention to them. Other local influential actors like traditional chiefs and NGOs do not have a mandate to be responsive and accountable, so when they voluntarily choose to be accountable to local people they are at best 'good despots' (Mills 2004, p. 36).

UN-REDD and Local Democracy

The UN-REDD Programme was launched in September 2008 and it is implemented by the three major UN agencies involved with environmental change and management—the UNDP, UNEP and FAO. The program's framework document maintains that the initiative 'grew out of requests from our respective governing bodies and rainforest countries to address issues related to forests and climate change' (UN-REDD 2008, p. 5). This implies that its legitimacy is dependent on its relationship with the governments that make up the UN. Consequently, UN-REDD license to carry out its activities in countries where it works is tied to the UN-REDD maintaining cordial relationship with the host country government.

The UN-REDD was created to 'assist forested developing countries and the international community to gain experience with various risk management formulae and payment structures'; in order to 'generate the requisite transfer flow of resources to significantly reduce global emissions from deforestation and forest degradation'; and test 'whether carefully structured payment structures and capacity support can create the incentives to ensure actual, lasting, achievable, reliable and measurable emission reductions while maintaining and improving the other ecosystem services forests provide' (UN-REDD 2008, p. 5).

To sum it up, the aim of the UN-REDD programme is to help to solve the climate change problem by incentivizing emissions reduction through money transfers from developed to developing countries. In response to this core underlying market logic in REDD+ initiatives, FERN (2011, p. 6) state that 'unless governance factors in forested countries are addressed as a priority, throwing money at the problem will do little to solve it'. This is because issues like mismanagement, corruption, and elite capture of REDD+ financial benefits for local communities, might derail achievement of lasting emissions reduction.

The UN-REDD programme is guided by 5 core principles (UN-REDD 2008, p. 7): a human-rights-based approach with attention to indigenous peoples' issues; gender equality; environmental sustainability; results-based management; and capacity development. The UN-REDD states that its 'rights-based and participatory approaches will also help ensure the rights of indigenous and forest-dwelling people are protected and the active involvement of local communities and relevant institutions in the design and implementation of REDD plans' (UN-REDD 2008, p. 7). It further states that its project execution strategy includes a 'REDD Dialogue' which would bring 'stakeholders together and ensure meaningful participation' (UN-REDD 2008, p. 11). Thus the UN-REDD favors the participatory stakeholder approach in its engagement with local groups and other actors including government and business.

This participatory stakeholder approach is inclusive of all actors interested in a resource, but it is not always attentive to power differential between weak and powerful actors, and ends up silencing the weak (Botes and van Rensburg 2000; Bulkeley and Mol 2003). The participatory stakeholder approach does not also differentiate between the democratic rights of citizens and interest of non-citizens like international business (Soma and Vatn 2014). Thus the voice of elected local leaders is easily dwarfed by that of more powerful actors.

On REDD+ compensatory payments, the framework document states that the UN-REDD will explore direct payments to persons with legal carbon rights, and indirect payment through central governments to local governments and local communities. The UN-REDD will test different compensatory payment mechanisms in order to determine the most optimal that will incentivize verifiable emissions reductions, while safeguarding local livelihoods. To ensure equitable payment distribution and reduce the risk of elite capture, the framework document refers to the need for 'strong democratic processes in local institutions' (UN-REDD 2008, p. 12). Principle 1, of the UN-REDD Social and Environmental Principles and Criteria document, states that the UN-REDD will 'apply norms of democratic governance, including those reflected in national commitments and Multilateral Agreements' (UN-REDD 2012a).

The 5 criteria that follow Principle 1 show how the UN-REDD would operationalize this principle in its project site based on key operators of transparency, accountability, legitimacy and participation. Criterion 4, which has to do with issues of inclusion and exclusion, states that the UN-REDD will 'ensure the full and effective participation of relevant stakeholders, in particular, indigenous peoples and other forest dependent communities, with special attention to the most vulnerable and marginalized groups' (UN-REDD 2012a, p. 4). The UN-REDD does not define 'participation' in its glossary of key terms, leaving it to the reader to infer that this likely implies inclusion of all stakeholders in deliberative processes to decide on REDD+ program goals and benefits. However, the UN-REDD does define 'relevant stakeholders' as:

> ...those groups that have a stake or interest in the forest and those that will be affected either negatively or positively by REDD+ activities. Relevant stakeholders include rights holders, those groups whose rights (human rights, customary or statutory rights, and/or collective rights) will be affected by REDD+ activities. These groups include relevant government agencies, formal and informal forest users, private sector entities, civil society, indigenous peoples and other forest dependent communities...

Missing from the above listing of relevant stakeholders are local government authorities. How would the UN-REDD strengthen local democracy if the immediate political representatives of local people are not prioritized as the principal local institutional stakeholder?

The use of the term stakeholders by the UN-REDD follows the trend by international organizations to opt for stakeholder democracy during project implementation. Stakeholder democracy is defined by MacDonald (2008) as governance arrangement that makes room for the private sector, non-state, and international organizations to be part of decision-making alongside national and sub-national government authorities (see also Backstrand 2006). However, stakeholder democracy dilutes citizens' rights over public resources. It also marginalizes citizens elected representatives like local government authorities because it places them on the same political standing with other stakeholders like business, who are often more powerful than local government authorities (Ribot 2003, 2004). Soma and Vatn (2014) show that when participatory processes prioritizes citizens' rights, outcomes favor public interests, but when participatory processes assume stakeholder equality, outcomes favor private interests.

The UN-REDD Social and Environmental Principles and Criteria document takes its democratic governance rhetoric from the UNDP's "A Guide to UNDP Democratic Governance Practice" (UNDP 2010). This document states that 'a major part of UNDP's assistance is geared towards advancing local democracy, focusing both on the core representative councils and assemblies and the mechanisms through which people can participate and hold their local government to account' (UNDP 2010, p. 58). Is this reflected in the UN-REDD ground activities in Nigeria-REDD, which it is funding to the sum of US$4 million?

How Nigeria-REDD Excluded Local Authorities During the Proposal Design Phase

The beginnings of the Nigeria-REDD program can be traced to the June 2008 Cross River State Stakeholders Summit on the Environment hosted by the governor, Senator Liyel Imoke (CRS 2008). The summit discussed how the state's forest resources could contribute more to revenue generation. Cross River State seeks new sources of revenue because it lost significant oil revenue of about US$115 million[4] per year (see Olubusoye and Oyedotun 2012), when Nigeria resolved its border dispute with Cameroon (see Konings 2011), and when the state resolved its boundary dispute with neighboring Akwa Ibom (see AKSG 2012).

The 2008 summit recommended that Cross River State ban logging and take up initiatives like REDD+ (CRS 2008; Oyebo et al. 2010). The governor proceeded to ban logging and expressed an interest in REDD+ (Oyebo et al. 2010). The governor also restructured the Cross River State Forestry Commission, appointing the well-known anti-logging activist Odigha Odigha, as the chairperson in 2009 (Filou 2010). Odigha Odigha won the prestigious Goldman Environmental Prize in 2003 for his anti-logging activism in Cross River State (Filou 2010). This was the first time a chairperson for a State Forestry Commission was appointed from the NGO sector. Odigha Odigha is well known to be a passionate advocate of REDD+.

In October 2009, Governor Imoke, and staff of the Forestry Commission attended the Katoomba XV meeting in Ghana (Oyebo et al. 2010). Katoomba is set up by Forest Trends, an international NGO, to promote payment for environmental services schemes like REDD+ (Forest Trends, the Katoomba Group and UNEP 2008). The governor invited experts from Katoomba to come and work with the State Forestry Commission to draft a REDD readiness plan idea note (R-PIN) for Cross River State (Oyebo et al. 2010). The R-PIN is a concept note submitted by a government authority to the UN-REDD to indicate an interest in implementing REDD+.

In November 2009, Governor Imoke asked the Nigerian Ministry of Environment to apply to UN-REDD for membership. In December 2009, he and a delegation from the Forestry Commission attended the UNFCCC conference of parties meeting in Copenhagen (COP 15), where he requested support for REDD+ in Nigeria (Oyebo et al. 2010). In January 2010, Katoomba visited pilot local communities in Cross River, and produced an R-PIN for Nigeria (FME 2011). By March 2010, Nigeria's membership request to UN-REDD was approved and in October 2010, a UN-REDD mission visited Nigeria (FME 2011).

Nigeria applied for membership of the UN-REDD in December 2009 (Oyebo et al. 2010) and its REDD readiness plan was approved for funding in October 2011 (FME 2011). Nigeria-REDD has a national program and a state level program with Cross River State as the pilot. At the national level, the Nigeria-REDD Secretariat is housed in the Department of Climate Change at the Ministry of Environment. This ministry works closely with the national advisory council on REDD and the national technical REDD committee. The advisory council is a policy making body, while the technical committee is a working group comprising of UN-REDD and Nigeria-REDD (national and state level) personnel. In addition, at the national level there is the REDD steering committee which is another working group for effective coordination of the work of the Department of Climate Change and the Cross River

---

[4]This is based on an exchange rate of 160.50 Naira to US$1 taking from www.xe.com as at September 2013; given that Cross River State Budget for 2013 was US$943 million (CRSG 2013), this revenue loss amounts to about 12% of the state budget for 2013 which would amount to a significant loss for any government.

State Forestry Commission (FME 2011). There is also a national civil society organizations' REDD forum, a platform for civil society to have a voice in Nigeria-REDD through the Department of Climate Change.

At the state level, the Cross River State government is the apex decision making body for REDD, so it is a member of the national advisory council on REDD. The state government directives are passed on to the Cross River Climate Change Council and to the forestry commission. The climate change council formulates state policy that is passed on to the state's Technical REDD Committee who translates this into a list of activities passed on to the forestry commission. The forestry commission is also influenced by decisions made at the Nigeria National Technical REDD Committee; commitments agreed to at the Nigeria REDD Programme Steering Committee meetings; inputs from the Climate Change Study Group at the University of Calabar; concerns from forest sector NGOs; by a small number of influential elites in local communities and traditional chiefs. REDD activities to be implemented by the forestry commission is carried out by Cross River State REDD team.

This institutional structure for REDD+ in Nigeria does not include local government authority but has multiple deliberative platforms. These multiple deliberative platforms are unavoidable, because they are part of federal and state government bureaucratic apparatus with formal accountability mechanisms. However, in a context like obtains in Nigeria where corruption is rife in government (see Olutola and Isaac 2016), these deliberative platforms create multiple level risks of elite capture of REDD+ benefits to local people. Where elite capture of local benefits from development projects is already a feature of the socio-political landscape, REDD+ has been observed to reinforce this dynamic (see Chomba et al. 2016).

Nigeria-REDD was designed with the active involvement of the UN-REDD Programme. The UNDP National Country Office (UNDP-NCO) is the UN in-country office primarily responsible for administering UN-REDD funds and supervising its use by the Nigeria-REDD team. The UNEP working out of its Nairobi office provides technical support on forest conservation and management in the Nigeria-REDD, while the FAO through its country office in Nigeria brings in expertise on developing national accounting systems for greenhouse gas inventories (FME 2011). The UNDP-NCO in an action plan strategy document for Nigeria maintains that deepening local democracy involves promoting 'stronger linkages and positive interaction between citizens and the first tier of government' (UNDP-NCO 2008, p. 9). The first tier of government here refers to the local government authorities. Would the UNDP-NCO rhetoric be evident in the Nigeria-REDD readiness proposal document?

Recall that by March 2010, Nigeria's membership request to UN-REDD was approved and it's REDD+ readiness plan was approved for funding in October 2011 (FME 2011). Part of the requirement for UN-REDD approval was for a broad based participatory consultative process to validate the REDD+ readiness plan. Thus on February 18, 2011, a REDD readiness proposal was presented to a participatory stakeholders' forum in Calabar chaired by Governor Imoke, for review and approval. The document was submitted to UN-REDD for consideration at its sixth policy board meeting in March 2011 (FME 2011). The board requested for revisions in the document (FME 2011; Global Witness 2011), which were effected and a second participatory stakeholders meeting was held in Calabar, in August 2011 to validate the revised document. Table 1 shows who attended the participatory consultative meetings in Calabar. The majority of the participants at the first meeting were from the forestry commission.

Local NGOs, customary authority and select members of local communities were also present, but there were no participants from elected local government authorities in Cross River State. Participants from local communities were mainly from Ekuri community because they have the largest community forest (330 km^2), and are the most active community forest group out of the 45 registered groups in Cross River State (Oyebo et al. 2010; UNDP 2012).

The select individuals from local communities invited to the participatory consultative meetings included farmer representatives, women representatives, and youth representatives. These were invited because the organizers considered them descriptively "typical of the larger class of persons whom they represent" (see Mansbridge 1999: 629). However, as descriptive representatives, they have no statutory mandate to report back to their local communities and cannot be sanctioned by their communities if they choose to be self-serving rather than stand for group interest. Hence, they are not democratically representative.

NGOs and traditional chiefs were the non-state local natural resources governance institutional actors invited to the consultative participatory meetings. Traditional chiefs base their legitimacy to represent their subjects on their shared cultural beliefs and norms reinforced through oral genealogical discourses. NGOs derive their legitimacy from securing social justice and poverty alleviation for local

Table 1 Participants at the Nigeria-REDD participatory consultative meetings

Institutions and groups	First meeting (2/18/11)	Second meeting (8/20/11)
Cross River State Forestry Commission	26	15
Local NGOs based in Cross River State	23	14
Participants from local communities (mainly Ekuri)	13	30
Media	8	2
Cross River State Governor's Office	6	0
International NGOs	6	1
Academics	6	4
Other Cross River State Government Agencies	5	0
Federal Ministry of Environment	2	0
National NGOs	2	0
Customary authority	2	6
Banks	2	1
Local Government Councils	0	0
Total	101	73

Source FME (2011)

people; that is the representative claims between NGOs and local people is based on their shared aspirations for social justice and poverty alleviation. This is why NGO legitimacy is often established and maintained through the use of iconic imagery of charismatic fauna and metaphors that promote the message that NGOs care about issues that their target groups are passionate about. This is best exemplified by WWF usage of the Giant Panda and their trademarked phrase 'for a living planet' in their official logo—this speaks loudly to the concerns of their predominantly European and American membership base.

In essence, NGOs and traditional chiefs are symbolic representatives of local people, this is because they are part of local governance arrangements, but are neither elected nor statutorily accountable to local people. NGOs and customary authority strengthen local democracy when they act as pressure groups on elected local authorities to be responsive. But they weaken local democracy when they replace these local substantive representatives. The absence of elected local governments from the Nigeria-REDD consultative process connotes that they were viewed as uninfluential, reflecting a lack of understanding of their role in strengthening local democracy in a representative democratic context as obtains in Nigeria. The participatory meetings therefore gave room for symbolic and descriptive representation but left out elected local government authorities—the substantive democratic representatives of local people.

Why Did Nigeria-REDD Readiness Proposal Exclude Local Government Authorities

In addition to being excluded from the participatory consultative meetings, the REDD+ readiness proposal document had very little to say about local government authorities compared with other tiers of government. The Nigeria REDD+ readiness proposal has 80 core activities planned and budgeted for, but only activity 3.2.3 about 'awareness raising for government officials, state legislators and local governments' refers to local government (FME 2011, p. 60). This is in contrast to numerous references to the national and Cross River State government, and Cross River State forestry commission. Why the blind spot with respect to elected local government? Based on the reading of all the documents cited above, the UN-REDD and Nigeria-REDD view 'stakeholder participation' as fulfilling UN-REDD (2008) and UNDP-NCO (2008) democratic governance rhetoric. However, not all participatory processes assure democratic representation of local communities in a forest governance regimen (Ribot 1996).

A UN-REDD representative who was asked about the exclusion of local government in the Nigeria-REDD readiness proposal said that

> UN-REDD cannot force countries to include the local level…there's a stakeholder engagement aspect looking to include local marginalized people…this include the free prior and informed consent process and concerns for indigenous people…there is also the participatory governance assessment process…to produce governance data…success depends on how civil

society actors would use it to hold government to account and how government would use it to do policy (UN-REDD Staff, personal communication, September 2012).

The response that countries cannot be forced to include the local level in their REDD+ readiness proposal reveals UN-REDD sensitivity to the sovereignty of its member governments. It also exposes the inability of the UN-REDD to use funding as leverage to get member governments to engage responsively with local people and with local democratic authorities that represent local socio-economic interests. Stating that the effectiveness of the UN-REDD governance model is dependent on civil society using it to hold government to account, shows UN-REDD confidence that NGOs can make governments responsive. The conviction that NGOs can make governments accountable and democratic reflects neoliberal thinking (see Mercer 2002; UN 2004; Sadoun 2007; Chorev 2013). However, NGOs in Nigeria are not always able to hold government accountable because they are co-opted as clients of government contracted to deliver social services (see Smith 2010; Fasakin 2011). NGOs can strengthen democracy when they act as watchdogs, and when they empower citizens and local authorities, acting in their symbolic role of inspiring society towards a just political order; they lose this virtue when they pursue government contracts to deliver social services (Banks et al. 2015).

The free prior and informed consent (FPIC) process and the participatory governance assessment (PGA) exercise are intended to capture local people's opinions as they are able to sincerely express those opinions in participatory settings (see UN-REDD 2012b; 2013b). These activities may strengthen the capacity of NGOs involved in their execution but do nothing to strengthen capacity of local democratic authorities mandated to act for local people.

The UN-REDD personnel also stated: *'strengthening local democratic governance is not the main priority of donors'* (UN-REDD Staff, personal communication, September 2012). Donors like the Norwegian government fund REDD+ because it is a cheaper means of reducing carbon emissions compared to regulating industries and restructuring their economy (Norwegian Government 2007; Eliasch 2008; UN 2008; Dyer et al. 2012). REDD+ allows donor countries to support global initiatives to mitigate climate change while allowing for business as usual (Cass 2012). Market and technocratic concerns dominate discourses on REDD+ showing it to be an ecological modernization project. As explained earlier in the chapter, ecological modernization projects address environmental problems without seeking to transform the economic processes that contribute to the problem (Hajer 1996; Baker 2007; Salleh 2010; Dyer et al. 2012; Roe et al. 2013; Nielsen 2014).

Conclusion

This chapter questions UN-REDD commitment to strengthen local democracy as a REDD+ safeguard, by examining the representation of local people in the consultative process that led to the US$4 million UN-REDD funded Nigeria-REDD. This is because REDD+ might exacerbate poverty in forest dependent communities by restricting forest access, and channeling compensatory payments to local communities through national government structures. The UN-REDD is convinced that strong local democratic institutions will ensure that compensation for local people are not captured by elites. However, UN-REDD is a global governance regime for which local people have limited information and are thus vulnerable to symbolic actions that may seem to protect local people, while advancing non-local interests. Overcoming this information gap, requires the UN-REDD to engage with local elected authorities and strengthen their capacity to respond to local needs including information needs.

The study finds that UN-REDD views the involvement of NGOs in the Nigeria-REDD consultative process as a sufficient indicator that their commitment to strengthen local democracy was being fulfilled (see also Nuesiri 2016). NGOs' claims to speak and stand for local people is based on shared socio-ecological and development discourses, which create affective linkages between NGOs and the aspirations of local people, for social justice and poverty alleviation. However, NGOs cannot be held to account by local people, so if they choose not to provide support to local communities to enable these communities protect their interests against other powerful REDD+ actors, they cannot be sanctioned by local people. Elected local authorities on the other hand are mandated to be responsive and accountable to local people; consequently, if they are involved in REDD+ design and implementation but are unresponsive to the concerns of their constituents, they can be sanctioned through the ballot box.

The study finds that the UN-REDD approved the democratically weak Nigeria-REDD readiness proposal because it trusts that local NGOs would be effective partners pushing for a responsive democratic governance agenda in Nigeria-REDD. It also approved Nigeria-REDD with its governance flaw because it was sensitive not to be seen as telling the Nigerian government authorities how to go about its business. Lastly, it approved Nigeria-REDD because it judged that its donors would not be too concerned with its weak attention to democracy as their focus is more on the market and technical aspects of REDD+. Thus the UN-REDD subscribed to symbolic politics in approving the Nigeria-REDD readiness proposal.

This finding that an international organization with a global governance agenda is subscribing to symbolic politics in its dealing with the public is not unique. As stated earlier, Baker (2007) maintains that the European Union (EU) subscribes to symbolic environmental politics in its commitment to sustainable development (a transformational paradigm), because EU operational strategies in dealing with environmental problems are informed by ecological modernization (a business as usual paradigm). How can this be changed?

In the case of the UN-REDD, it should not engage at the local level with only NGOs, but also with elected local government authorities. Where democratic local governments do not exist, the UN-REDD should seek to move local governance arrangements in a democratic direction. The UN-REDD should carry out its FPIC and PGA in full partnership with local government authorities, and invite them to participate in UN-REDD board meetings. If UN-REDD engages with local governments this way, it would make their activities more transparent and increase downward accountability to the local level (see also Nuesiri 2016).

Strong local democracy is the basis for strong democratic governance; the UN guidance note on democracy states that "strong and effective local democratic institutions are an underlying basis for a healthy democracy... are more accessible for citizens to question local officials...present their interests and concerns and resolve their disputes...and can be an arena for attracting new political actors, including women and young people" (UNSG 2009, p. 8). Therefore, to ensure that climate governance and local democracy are synergistic and not dissonant, the UN-REDD must find innovative ways to strengthen local democratic processes, despite the very real constraints of respecting the sovereignty of its member states and the low level of interest among its donors in the objective of achieving democracy.

Acknowledgements This chapter is based on research carried out for the Responsive Forest Governance Initiative (RFGI), funded by the Swedish International Development Agency (SIDA), and executed by the Council for the Development of Social Science Research in Africa (CODESRIA) Dakar, the International Union for the Conservation of Nature (IUCN), and the University of Illinois Urbana Champaign (U of I). The author thanks all RFGI colleagues for the rich conceptual discussions that informs the theoretical position of this Chapter, notably Jesse Ribot, James Murombedzi, and Gretchen Walters. Thanks also to my wife and daughters for their support.

References

Accra Caucus (2013) REDD+ safeguards: more than good intentions? Case Studies from the Accra Caucus. Report of the Accra Caucus on forests and climate change. Available from: http://www.fern.org/sites/fern.org/files/Accra%20report%20vol%203_eng_final.pdf. Accessed 20 Jan 2016

AKSG (2012) 76 Oil wells—how cross river state went to court and lost. Akwa Ibom State Government. (http://www.africanspotlight.com/2012/07/23/76-oil-wells-how-cross-river-state-went-to-court-and-lost/). 25 Nov 2013

Backstrand K (2006) Democratizing global environmental governance? Stakeholder democracy after the World Summit on Sustainable Development. Eur J Int Relat 12(4):467–498

Baker S (2007) Sustainable development as symbolic commitment: declaratory politics and the seductive appeal of ecological modernisation in the European Union. Environ Polit 16(2):297–317

Banks N, Hulme D, Edwards M (2015) NGOs, States, and Donors Revisited: Still Too Close for Comfort? World Dev 66:707–718. https://doi.org/10.1016/j.worlddev.2014.09.028

Beymer-Farris BA, Bassett TJ (2012) The REDD menace: resurgent protectionism in Tanzania's mangrove forests. Glob Environ Change 22(2):332–341

Bluhdorn I (2007) Sustaining the unsustainable: symbolic politics and the politics of simulation. Environ Polit 16(2):251–275

Botes L, van Rensburg D (2000) Community participation in development: nine plagues and twelve commandments. Commun Dev J 35(1):41–58

Brysk A (1995) "Hearts and minds": bringing symbolic politics back in. Polity 27(4):559–585

Bulkeley H, Mol APJ (2003) Participation and environmental governance: consensus, ambivalence and debate. Environ Values 12(2):143–154

Cass LR (2012) The symbolism of environmental policy: foreign policy commitments as signaling tools. In: Harris PG (ed) Environmental change and foreign policy: theory and practice. Routledge, pp 41–56

Chomba S, Kariuki J, Lund JF, Sinclair F (2016) Roots of inequity: how the implementation of REDD+ reinforces past injustices. Land Use Policy 50:202–213

Chorev N (2013) Restructuring neoliberalism at the World Health Organization. Rev Int Polit Econ 20(4):627–666

CIF, FCPF and UN-REDD (2010) Enhancing cooperation and coherence among multilateral REDD+ institutions to support REDD+ activities. Climate Investment Funds, Washington DC

Coleman S (2005) New mediation and direct representation: reconceptualizing representation in the digital age. New Media & Society 7(2):177–198

Corbera E, Schroeder H (2011) Governing and implementing REDD+. Environ Sci Policy 14(2): 89–99

CRS (2008) Communique of the stakeholders' summit on the environment organized by the government of cross river state. Calabar, Nigeria. (http://tropicalforestgroup.org/wp-content/uploads/2012/10/crsoutcomesdOC.pdf). 25 Nov 2013

CRSG (2013) Consolidated financial statement. (http://www.crossriverstate.gov.ng/index.php?option=com_phocadownload&view=category&id=12:2013-budget). 6 Feb 2015

Dahl RA (1989) Democracy and its critics. Yale University Press

Davenport C (2007) State repression and the domestic democratic peace. Cambridge University Press

Dyer N, Counsell S, Cravatte J (2012) Rainforest Roulette? Why creating a forest carbon offset market is a risky bet for REDD. Climate and forests policy brief. Rainforest Foundation, London

Edelman M (1985) The symbolic uses of politics. University of Illinois Press

Eliasch J (2008) The Eliasch review—climate change: financing global forests. Report from the office of climate change. Department of Energy and Climate Change, London. Available from: https://www.gov.uk/government/uploads/system/uploads/attachment_data/file/228833/9780108507632.pdf. Accessed 20 Jan 2016

Fasakin A (2011) Non-governmental organisations and civil corruption in Nigeria's public space. Chapter presented at the international conference on corruption, governance and development in Nigeria. Aminu Kano Centre for Democratic Research and Training. Available from: https://www.academia.edu/2400888/NGOs_and_Civil_Corruption_in_Nigerias_Public_Sphere. Accessed 20 Jan 2016

FERN (2011) REDD+ and carbon markets: 10 myths exploded. Briefing Chapter by FERN Brussels, Greenpeace International Amsterdam, Rainforest Foundation UK and Friends of the Earth USA. Available from: http://www.fern.org/sites/fern.org/files/10%20myths%20exploded_new.pdf. Accessed 20 Jan 2016

Filou E (2010) Odigha Odigha: speaking truth to power. (http://www.ecosystemmarketplace.com/pages/dynamic/article.page.php?page_id=7497&section=news_articles&eod=1). 25 Nov 2013

FME (2011) Nigeria's REDD+ readiness programme (2012–2015). Federal Ministry of Environment, Abuja. Available from: http://mptf.undp.org/document/download/10974. Accessed 20 Jan 2016

Forest Trends, the Katoomba Group and UNEP (2008) Payments for ecosystem services: getting started—a primer. Forest Trends, Washington, DC

Fraser N (2005) Reframing justice in a globalizing world. New Left Rev 36:69–88

Frechette A., de Bresser M, Hofstede R (2014) External Evaluation of the United Nations Collaborative Programme on Reducing Emissions from Deforestation and Forest Degradation in Developing Countries (the UN-REDD Programme) Volume I—Final Report. http://www.fao.org/fileadmin/user_upload/oed/docs/UN-REDD%20Global%20Evaluation%20Final%20Report.pdf. Accessed 25 April 2018

Global Witness (2011) Review of Nigeria's NPD submitted to the 7th policy board meeting of the UN-REDD programme. Available from: (https://www.globalwitness.org/sites/default/files/nigeria%20npd.pdf). Accessed 31 May 2016

Grunebaum J (1981) What ought the representative represent? In: Bowie N (ed) Ethical issues and government. Temple University Press, pp 54–67

Hajer MA (1996) Ecological modernisation as cultural politics. In: Lash S, Szerszynski B, Wynne B (eds) Risk, environment and modernity: towards a new ecology. SAGE Publications, pp 246–268

Hogan J, Littlemore R (2009) Climate cover up: the crusade to deny global warming. Greystone Books

Kateb G (1981) The moral distinctiveness of representative democracy. Ethics 91(3):357–374

Keck ME, Sikkink K (1999) Transnational advocacy networks in international and regional politics. Int Soc Sci J 51(159):89–101

Konings P (2011) Settling border conflicts in africa peacefully: lessons learned from the Bakassi dispute between Cameroon and Nigeria. In: Abbink J, de Bruijn M (eds) Law and politics in Africa: mediating conflict and reshaping the state. Leiden, Brill, p 191

Larsen J, Larsen K, Herndon W, Mohan S (2016) Taking stock: progress toward meeting US climate goals. Rhodium Group Report, New York, NY

Lombardo E, Meier P (2014) The symbolic representation of gender: a discursive approach. Ashgate Publishing Company

Macdonald T (2008) Global stakeholder democracy: power and representation beyond liberal states. Oxford University Press, Oxford

Manin B, Przeworski A, Stokes S (1999) Introduction. In: Przeworski A, Stokes S, Manin B (eds) Democracy, accountability and representation. Cambridge University Press, pp 1–26

Mansbridge J (1999) Should blacks represent blacks and women represent women? A contingent "yes". J Polit 61(3):628–657

Matten D (2003) Symbolic politics in environmental regulation: corporate strategic responses. Bus Strategy Environ 12(4):215–226

Mercer C (2002) NGOs, civil society and democratization: a critical review of the literature. Prog Dev Stud 2(1):5–22

Mill JS (2004) Considerations on representative government. Penn State Electronic Classics Series Publication. Available from: http://www2.hn.psu.edu/faculty/jmanis/jsmill/considerations.pdf. Accessed 20 Jan 2016

Miller HT (2012) Governing narratives: symbolic politics and policy change. University of Alabama Press

Montanaro L (2012) The democratic legitimacy of self-appointed representatives. J Polit 74(4):1094–1107

Newig J (2007) Symbolic environmental legislation and societal self-deception. Environ Polit 16(2):276–296

Nielsen TD (2014) The role of discourses in governing forests to combat climate change. Int Environ Agreements Polit Law Econ 14(3):265–280

Norwegian Government (2007) Norwegian climate policy. Norwegian Ministry of the Environment, Oslo. Available from: https://www.regjeringen.no/contentassets/c215be6cd2314c7b9b64755d629ae5ff/en-gb/pdfs/stm200620070034000en_pdfs.pdf. Accessed 20 Jan 2016

Nuesiri EO (2016) Deepening local democracy for a more just global governance regime. ISS Colloquium Paper No. 30, International Institute of Social Studies, The Hague, Netherlands

Ntsebeza L (2005) Democracy compromised: chiefs and the politics of the land in South Africa. Brill, Leiden, Boston

Olubusoye OE, Oyedotun TDT (2012) Quantitative evaluation of revenue allocation to states and local governments in Nigeria (1999–2008). Eur Sci J 8(3):224–243

Olutola OF, Isaac OO (2016) Legislative Corruption and the Challenges of Democratic Consolidation in Nigeria. Appl Sci Rep 14(3):230–236. https://doi.org/10.15192/PSCP.ASR.2016.14.3.230236

Oyebo M, Bisong F, Morakinyo T (2010) A preliminary assessment of the context for REDD in Nigeria. Federal Ministry of Environment, Abuja. Available from: http://www.unredd.net/index.php?option=com_docman&task=doc_download&gid=4129&Itemid=53. Accessed 20 Jan 2016

Pitkin HF (1967) The concept of representation. University of California, Berkeley

Rehfeld A (2006) Towards a general theory of political representation. J Polit 68(1):1–21

Rehfeld A (2011) The concepts of representation. Am Polit Sci Rev 105(3):631–641

Ribot JC (1996) Participation without representation: chiefs, councils and forestry law in the West African Sahel. Cultural Survival Q 20(1):40–44

Ribot JC (2003) Democratic decentralisation of natural resources: institutional choice and discretionary power transfers in Sub-Saharan Africa. Public Adm Dev 23(1):53–65

Ribot JC (2004) Waiting for democracy: the politics of choice in natural resource decentralization. World Resources Institute, Washington DC

Roe S, Streck C, Pritchard L, Costenbader J (2013) Safeguards in REDD+ and forest carbon standards: a review of social, environmental and procedural concepts and application. Climate Focus, Amsterdam

Sadoun B (2007) Political space for non-governmental organizations in United Nations World Summit processes. Civil Society and Social Movements Programme Chapter Number 29. United Nations Research Institute for Social Development, Geneva. Available from: http://www.unrisd.org/80256B3C005BCCF9/search/119D7568A3373C47C12572C900444EFF? OpenDocument. Accessed 20 Jan 2016

Salleh A (2010) Climate strategy: making the choice between ecological modernisation or living well. J Aust Polit Econ 66:118–143

Saward M (2006) The representative claim. Contemp Polit Theory 5(3):297–318

Saward M (2008) Representation and democracy: revisions and possibilities. Sociol Compass 2:1000–1013

Silveira S (2004) The American environmental movement: surviving through diversity. Boston College Environ Aff Law Rev 28(2):497–532

Sisk TD (ed) (2001) Democracy at the local level: the international IDEA handbook on participation, representation, conflict management, and governance. International Institute for Democracy and Electoral Assistance (International IDEA), Stockholm, Sweden

Smith DJ (2010) Corruption, NGOs, and development in Nigeria. Third World Q 31(2):1–13

Soma K, Vatn A (2014) Representing the common goods—stakeholders vs. citizens. Land Use Policy 41:325–333

Stavins RN (1998) What can we learn from the grand policy experiment: lessons from SO_2 allowance trading. J Econ Perspect 12(3):69–88

UN (2004) We the Peoples: civil society, the United Nations and global governance. Report of the Panel of Eminent Persons on United Nations–Civil Society Relations. United Nations, New York. Available from: http://www.un-ngls.org/orf/Final%20report%20-%20HLP.doc. Accessed 20 Jan 2016]

UN (2008) 'REDD'-letter day for forests: United Nations, Norway unite to combat climate change from deforestation, spearheading new programme. UN Press Release ENV/DEV/1005. Available from: http://www.un.org/press/en/2008/envdev1005.doc.htm. Accessed 20 Jan 2016

UNDP (2012) Ekuri initiative, Nigeria. Equator initiative case study series. UNDP, New York. Available from: (http://www.ng.undp.org/publications/Equator-Award_Ekuri-initiative.pdf). Accessed 31 May 2016

UNDP (2010) A guide to UNDP democratic governance practice. UNDP Oslo Governance Centre, Oslo, Norway

UNDP-NCO (2008) Country programme action plan between the Federal Republic of Nigeria and the United Nations Development Programme—Nigeria. UNDP Nigeria, Abuja

UN-REDD (2008). UN collaborative programme on reducing emissions from deforestation and forest degradation in developing countries (UN-REDD): An FAO, UNDP, UNEP Framework Document. UN-REDD Programme, Geneva

UN-REDD (2012a) UN-REDD programme social and environmental principles and criteria. UN-REDD Programme, Geneva

UN-REDD (2012b) Abridged report: participatory governance assessment (PGA) pilot research in the Cross River State, Nigeria. UN-REDD Programme, Geneva

UN-REDD (2013a) Guidelines on free, prior and informed consent. UN-REDD Programme, Geneva

UN-REDD (2013b) Workshop report: participatory governance assessments consultative workshop Calabar—Nigeria. UN-REDD Programme, Geneva

UNSG (2009) Guidance note of the secretary-general on democracy. Available from: http://www.un.org/en/globalissues/democracy/pdfs/FINAL%20Guidance%20Note%20on%20Democracy.pdf. Accessed 20 Jan 2016

Urbinati N, Warren M (2008) The concept of representation in contemporary democratic theory. Annu Rev Polit Sci 11:387–412

Wedeen L (1998) Acting "as if": symbolic politics and social control in Syria. Comp Stud Soc Hist 40(3):503–523

Wintrobe R (1998) The political economy of dictatorship. Cambridge University Press

Chapter 11
Sea Level Rise and Social Justice: The Social Construction of Climate Change Driven Migrations

Elizabeth Marino

Abstract One outcome of climate change will be sea level rise. Sea level rise, and subsequent flooding, may cause displacement of certain people and communities. Social science research has argued that disaster outcomes, such as displacement by flooding, are socially constructed—that is, they are the outcomes of decisions made about where to develop, who the state protects, and how communities recover following an environmental hazard. This chapter addresses the idea that sea level rise is intimately linked to questions of social justice using three case studies. First the chapter investigates vulnerability as an outcome of colonization practices in Alaska. Next, the chapter addresses the impacts of cost-benefit analysis for beach nourishment on coastal populations. Finally, the chapter will look at the Isle de Jean Charles example from Louisiana to understand how cost-benefit analysis impacts levee protection decisions. Ultimately the chapter will argue that the suffering caused by sea level rise is a social construct, as well as an outcome of ecological shift. Here we see that habitual marginalization and economic and political systems of disenfranchisement render certain populations invisible or "less valuable" to protect, and that this, in turn, perpetuates cycles of vulnerability under climate change regimes.

Introduction

Climate change creates new ecological baselines. The most striking of these, and the changes that often garner the most attention in the popular press (Hartman 2010; Marino 2012; Farbotko and Lazrus 2012), are changing coastal conditions and shrinking coastlines due to sea level rise (Nichols and Cazenave 2010). We understand climate change driven, global sea level rise to be linked primarily to ocean expansion (Rhamstorf 2007), and melting fresh water from Greenland

E. Marino (✉)
Social Science and Sustainability, OSU—Cascades, 1103, NW,
Stannium Road, Bend, OR 97703, USA
e-mail: elizabethmarino@osucascades.edu

(Rignot et al. 2011) and Antarctica (Deconto and Pollard 2016)—and separate from decadal changes in regional sea levels linked to winds and other regional drivers (Church et al. 2013). In other words, as the ocean warms and expands; and, as glaciers melt, and previously frozen water moves into the ocean system, we can expect significant changes in the coastlines of the world. Conservative estimates indicate that by 2100 there will be between 35 and 74 cm rise in sea level (Hinkel et al. 2014); however, other studies suggest there could be 1 m or more rise in global sea level rise by the same year (Nicholls and Cazenave 2010, Pfeffer et al. 2008). Without a dramatic change in green house gas emissions these rates are expected to accelerate after 2100 (Church et al. 2013).

An important fact to keep in mind as we consider sea level rise is that coastal communities are often population dense and their population growth is accelerating compared to non-coastal communities and locations (Neumann et al. 2015). The majority of megacities around the world are located within the low-elevation coastal zone (LECZ), defined as the "contiguous and hydrologically connected zone of land along the coast and below 10 m of elevation" (Neumann et al. 2015). Urban areas in LECZs are also growing at higher rates than urban areas outside of coastal areas. In 2000, 625 million people lived in LECZs, the majority of which were in less developed countries (Neumann et al. 2015).

Sea level rise is—in many ways, rightly—linked in the global imagination to the inundation of these megacities and towns that lie along the coast, or to the inundation of island-nations, all of which are surrounded by an encroaching, ever-threatening ocean (Farbotko and Lazrus 2012). Inundation is often conceived of as the driver of climate change displacements, and is central to a powerful narrative that is often used to convey the seriousness and scale that climate change impacts could have on the world's population.

Indeed, the scale of impacts of sea level rise and other climate change outcomes are driving calls for action. Researchers and the global community are increasingly vocal about the links between climate change impacts and displacement of people and communities (Bettini 2013). The fifth IPCC report reported that anthropogenic climate change is currently pushing some communities to relocate in response to a host of factors ranging from sea level rise and more extreme weather, to livelihood disruption linked to drought and other climatological burdens, and that this trend towards climate change driven migration is expanding (Adamo and de Sherbinin 2011). The Stern review, in alignment with Norman Myers' early estimates (Myers 1993), predicted that, by 2050, as many as 150–200 million migrants may be pushed out of their current locations because of climate change (2006); and the International Organization on Migration has pushed that number even higher, claiming that there may be as many as 1 billion environmental migrants by the year 2050 (Oli 2008).

These risks and threats give rise to a striking and harrowing vision of the future in which individuals and communities will be forced from their homes at alarming rates and in alarming numbers. This dystopian vision has struck some social critics to identify the climate refugee narratives as apocalyptic—fanning xenophobic visions of hordes of people trying to "get in" to developed countries (Bettini 2013;

Marino 2015; Farbotko and Lazrus 2012). Some researchers question the veracity of the estimates themselves, claiming that future climate refugee numbers are based on insufficient data and lack methodological rigor (Gemenne 2011); but many are concerned with the narratives themselves as a mechanism of "othering". Giovanni Bettini points out that titles of articles and popular pieces such as "The Human Tide" (Baird 1992), and "The Human Tsunami" (Knight 2009) are symptomatic of the apocalyptic, "othering" of potential climate change migrants (2013). Ultimately, the critique is that hyperbolization and dystopian visions of climate change migrants depoliticizes displacement narratives by masking power dynamics and simplifies complex stories and histories of class, race, political power and pre-existent social drivers which interact with climate outcomes to cause displacement (Marino and Ribot 2012; Marino 2015).

Amid the noise of these public and scientific narratives—including real and serious stories about climate change outcomes and risks, and apocalypitic dystopian narratives—lies a serious social science question: to what extent are climate change displacement risks the outcome of social injustice? This chapter addresses these issues using three case studies from the United States. While all three of these cases are located in the US, the social dynamics of decision-making and vulnerability creation will likely have applicability across the globe. Indeed, the cases I am highlighting here have been and are being compared to case studies elsewhere (Marino and Lazrus 2015), and help to demonstrate that vulnerability and vulnerable communities are also constructed within the relative wealth and prosperity of a "northern" economy. The IPCC notes that, "socially and geographically disadvantaged people exposed to persistent inequalities at the intersection of various dimensions of discrimination based on gender, age, race, class, caste, Indigeneity, and (dis)ability are particularly negatively affected by climate change and climate-related hazards". This chapter will help to unpack why this is so.

This chapter will first discuss the social construction of disasters and the implications of disaster theory in the context of climate change displacement and relocation. Next I summarize the interactions among colonial legacies and relocation stories on the coast of Alaska. Then I will look at issues of decision-making regarding beach nourishment along the Eastern Seaboard of the United States. Next, I will look at decisions regarding coastal protection in Louisiana and the decisions and costs to protect people at home, or move them out of harm's way. Finally, I will discuss the similarities among these sections and how they can be instructive as we attempt to create and sustain solutions for communities who face risk—be they social or environmental.

The Social Construction of Suffering Caused by Sea Level Rise

One emerging aspect of the climate change and migration debate has been an attempt to define terminology and create a typology of migrants. This has proven to be difficult. First, the movement of people itself exists on a scale from voluntary migration to forced displacement. Many have noted the political and legal importance of pointing out when, and in what circumstances, displacement is "forced" and linked to the dispossession of land and resources (Oliver-Smith 2009). For example, there are currently mechanisms being developed (The Warsaw International Mechanism on Loss and Damage) that may repay communities displaced by climate change a portion of what they have lost. In this vein, it is critical to identify climate change-driven migrations as displacement in order to underscore and justify the legal rights displaced peoples may be entitled to following the act of displacement. Identifying displacement as a form of injustice is fitting in no small part because it is often the communities who have contributed the least to green house gas emissions which are the most vulnerable to climate change impacts (Campbell-Lendrum and Corvalán 2007), including displacement.

However, some researchers have pointed out that economic migrations are hard to untangle from climate change migrations on the ground. While ecological changes are often part of migrating decisions, to date, ecological shift is most typically a single factor in the suite of push and pull factors which drive individuals and families to make migration decisions (Warner 2010). Additionally, climate change impacts and risks can trigger disinvestment in infrastructure and economic sectors—leading individuals to decide to migrate as a result of disinvestment (Marino and Lazrus 2015). In cases where there is significant disinvestment in a community because of climate change risks, are people who move economic migrants or victims of displacement due to climate change impacts? For the rest of this chapter, people who move and whose decision to move is in part an outcome of climate change impacts will be identified as climate change migrants. This designation is applicable to the wide variety of experiences that climate change migrants might have; yet we will see that certain populations are still more at risk of climate migration than others.

The complexities of typology highlight what social scientists of natural disasters have demonstrated for the last four decades—that ecological events enter into stratified societies and social decision-making—creating complex, socio-ecological systems, within which people are pushed to migrate (Marino and Ribot 2012). In the anthropological literature, we understand disasters such as floods and drought to be socially constructed (Oliver-Smith and Hoffman 1999). Most generally, the social construction of disaster and risk, including climate change risks, can be understood as the sum total of decisions that organize and orient human networks and materials and into which an ecological event or series of events enter. If an ecological event, such as high water, enters a space which is not occupied by people, it is not considered (by most, see Purser et al. 1995) to be a natural disaster. If an ecological event, such as high water, enters a space, but does not disrupt social life, it is not considered a disaster. If an ecological event or series of events, such as the slow erosion of a river bank, pushes individuals to move, but without interrupting social life or habits, then that event is not considered a disaster. The word disaster always connotes a disruption from social life; but, and more importantly, human decision-making creates spaces and power hierarchies which render some peoples' social lives more vulnerable to disruption than others when an ecological disturbance (or hazard) occurs. In the end, then, the word disaster masks history as nature; and social decisions as the whims of fate.

In other words, the decisions humans made and continue to make regarding where to build cities, who to protect with sea walls, how to retreat, and how to mitigate changing ecological conditions, are key in determining whether or not a disaster occurs. Using these insights to discuss sea level rise and climate change-driven migrations we might say that while sea levels rise will occur and humans will have to move—it is entirely a social choice, not an ecological one, whether or not there will be human suffering accompanying those changes. If, for example, communities threatened with sea level rise were rebuilt in ways that respected pre-existent traditions, empowered community decision-making, and *improved* infrastructure, suffering could be lowered even as the migration occurred. Would, then, this be a disaster?

We can see these social and political characteristics of climate change driven migration playing out in case studies around the United States and around the world. In places where climate change impacts are already creating pressures on individuals and communities to relocate, the social and political contexts in which climate change outcomes enter are fundamental to understanding the experience of communities, and the constraints on these communities' adaptation options. First, socio-political decisions and injustices of the past fundamentally create some of the most visible examples of climate change driven migration today, as exemplified by cases of Indigenous peoples being forced to relocate in Alaska. In other cases, climate change and sea level rise mitigation decisions, as an outcome of cost-benefit analyses, render certain communities "worthy" of protection and others less so. These decisions are instructive in understanding the options of local, regional, state, and international communities in the larger goal to protect people and places from harm.

The Colonial Legacy of Climate Change Migrations in Alaska

In Alaska, the effects of climate change are already tangible. The Arctic experiences something called Polar Amplification, which is an extreme warming around the poles compared to the rest of the world, during periods of warming trends. To date, the Arctic is warming at twice the rate of the rest of the world (Screen and Simmonds 2010). While the goals of the Paris agreement are to limit warming globally to 2 °C, models for the Arctic region place warming between 5 and 7 °C by the end of the century (Kattsov et al. 2005).

Polar Amplification is complex (Serreze and Barry 2011); but to characterize it most simply, changes in Arctic sea ice impact the overall power of the earth's albedo, with localized outcomes. The Earth's albedo is the reflective power of the earth. In the Arctic, as warming occurs, Arctic sea ice decreases. When Arctic sea ice decreases, heat and radiation from the sun stops being reflected back into the atmosphere (albedo) and instead is absorbed. This has the regional impact of increasing temperatures more quickly and more profoundly in the Arctic than in other areas of the globe during warming trends.

Impacts of Polar Amplification have led to changes in many parts of the Arctic climate system—some of which impose serious pressures on communities. These include permafrost thaw, increased storminess in some regions, and the delayed forming of pack ice in the ocean (Larsen et al. 2014). In tandem these changes can lead to flooding—and as these changes persist, habitual flooding can become a new ecological baseline. In the Arctic, flooding and erosion impact 86 percent of Alaska Native communities (GAO 2009).

Some Arctic communities have begun to experience habitual flooding extreme enough to render current locations uninhabitable for the future. The communities of Shishmaref, Kivalina, and Newtok, Alaska, for example, have been at the forefront of efforts to adapt to climate change impacts through relocation. Their community-led efforts have been discussed in a number of research publications (Bronen 2008, 2010, 2011; Bronen and Chapin 2013; Bronen et al. 2017; Maldonado et al. 2013; Marino 2012; Shearer 2011) and all conclude that climatic impacts interact with policy obstacles and other social phenomena to create risky conditions for communities. In particular, scholars have argued that there is no governance framework to facilitate an organized relocation and have called for governance frameworks to be developed which are rooted in human rights and the principals of self-determination (Bronen and Chapin 2013; Bronen et al. 2017).

Less discussed in these publications are the impact that histories and legacies of colonial institutions have on creating risk in these situations. My work in Shishmaref, Alaska began with a friend and colleague named Tony Weyiouanna pointing out to me frankly, "no one is asking how we got here in the first place." In Shishmaref, traditional livelihoods and lifestyles were predicated on the flexibility to move with seasonal changes and changes in ecological conditions. Infrastructure and housing used, at that time, were well suited to highly mobile lifestyles. It appears that in the past, a sophisticated interaction among social habits and local ecologies created high resilience to changing coastal conditions.

At around the turn of the 20th Century, colonial institutions, Western market economies, and forced schooling policies pushed into western Alaska. Often, a first priority of colonial leaders was to sedentarize smaller communities into centralized locations (Ducker 1996) and build immobile infrastructure (school, church, post office, and, later, stick houses). These locations were often located in spring subsistence hunting sites and, therefore, were built in places that allowed for people to continue, though altered, subsistence practices that secured both mundane and sacred relationships with the landscape and animals. The immobile infrastructure that was part and parcel of colonization, however, could become exposed to high water when and if coastal conditions fluctuated. Indigenous communities, whose traditional infrastructure could be moved when ecological conditions changed, were now located in buildings that did not move. Community members, therefore, were exposed to what was, for the first time, the social interruption or disaster, which we call flooding (Marino 2015).

The link between contemporary climate change driven flooding and colonial legacies, therefore, lies not just in the historical consequences that removed high mobility as an adaptation strategy of the past—but in colonial decision-making which rendered traditional adaptation strategies inaccessible without creating new and adequate adaptation strategies to replace them. In other words, if participation in immobile infrastructure and state-driven institutions were a product of colonization—then the cost of flooding today is, and should be, incurred by the state. To date, however, there has not been money allocated through either the state of Alaska or the federal government of the United States that is sufficient to ensure the relocation of any community in Alaska exposed to habitual flooding. The point of directing attention to how colonial legacies construct vulnerability is not to suggest a hierarchy of deservedness—i.e. because Indigenous people have suffered greatly they deserve state intervention (though that may be true). Here, I point to the histories of colonization to demonstrate that vulnerability was *created* by colonization. The ideology and decision-making of the state, as it confiscated land and resources from Indigenous peoples and intentionally dismantled traditional economies and life-ways, created socio-geographical spaces that are more vulnerable to ecological shift and climate change today, and simultaneously these are also communities less likely to have the political power to harness state funds in order to adapt. These risks to Alaska Native communities are all the more morally problematic because a large flood that occurred today could potentially dispossess the community from their homeland—that very definitive act of colonization. In my work in Shishmaref, I learned that a leading fear of community members was to be relocated outside of traditional territory in the wake of a large storm (Marino 2015).

In the Alaska cases, the decisions made by colonial actors in the past to force schooling and other traumas into Alaska Native communities is a social injustice that generated risk. Likewise, the lack of political will to budget money to relocate communities in an organized and community-led way is, today, a social injustice that is rendering risk. Climate change enters into social situations in which certain communities are more vulnerable linked to historical contexts and through which certain communities remain vulnerable linked to the power structures which continue the colonial project. This case is not meant to suggest that colonial histories are the only histories which create vulnerability—but because we can see how colonial decision-making is part of that which creates vulnerability to climate change today, it suggests that history, not the size of the storm, are the roots of suffering caused by sea level rise.

Beach Nourishment and Cost-Benefit Analysis: The Price of Protection

Issues of class are also persistent drivers of climate change outcomes, including migration. As demonstrated in Venice, Italy (Nosengo 2003), as well as the islands being built in the south China sea (Watkins 2015), defending structures against water inundation may be expensive and complex, but there is much that can be done. While the two examples above are extreme examples of protecting or building land to prevent inundation—defense against erosion is relatively common. Defending against sea water inundation is possible with both hard structures, such as sea walls and levies, and soft engineering such as beach nourishment. Both are expensive. Beach nourishment in the US cost $787 million from 1995 to 2002 (Gopalakrishnan et al. 2011). Most researchers conclude that these costs cannot be incurred in every place that is threatened with flooding and sea level rise. Thus, politicians and other decision-makers have to choose where to invest in protection. The fundamental tool in deciding where beach protection and defense against sea level rise should occur is cost-benefit analysis. For example, if the costs of beach nourishment are lower than the loss of what the beach protects, or if the beach itself is an economic asset, then beach nourishment is seen as a good investment. Because of this, places with higher property values and tourist locations are more likely to be protected than other places (Smith et al. 2009).

There are, however, social justice concerns when applying these kinds of analyses. First, it is widely accepted that urban areas are prioritized for coastal management and defense against sea level rise (Cooper and McKenna 2008). Along the US eastern sea board, this suggests that large urban areas will be targeted for protection, while communities from more rural areas will face migration and retreat from rising sea levels. One can imagine a future in which there are none-to-few rural coastal communities in the US that are not tourist locations, if the resources of protection are spent on increasingly expensive defenses for cities. In rural Maryland, for example, we see scores of small communities suffering under flooding and erosion without the benefits of protection either by hard infrastructure or beach nourishment (Fiske and Marino in press). While there may be an argument for protecting urban environments because of their large populations—the reverse of "protection", namely, "sacrifice" is never called out explicitly. Who are we deciding not to protect? Are these communities consistently poor and vulnerable populations? What are the consequences of these decisions?

A second concern for a cost-benefit analysis model of beach nourishment is the focus on protecting tourism locations (Houston et al. 1996). Beach nourishment is considered economically viable if the value of the beach on the local economy is greater than the cost of transporting and laying down new sand. These calculations only work if the beach is an economically generating resource, which it is in tourist hubs. In fact, a brochure put out by the Army Corps of Engineers on beach nourishment specifically identifies tourism as a central concern and justification for beach nourishment efforts (US Army Corps of Engineers 2007). However, tourism industries often drive up costs of living and can displace less wealthy community members away from the waterfront and towards more inland locations or less expensive coastal towns (Thompson et al. 2016). Here, then, protection from sea level rise is predicated on the same social forces (tourism and gentrification) which can drive out-migration of more poor individuals, families, and communities, into spaces which are not considered economically-viable to protect.

Finally, all federal beach protection in the United States is conducted through the Army Corps of Engineers, an institution dubbed by some scholars as the "political grandfather of cost-benefit analysis in the United States" (Persky 2001). Economists have noted that the deployment of cost-benefit analyses gives an illusion of rationality, when in fact, decisions on whether to carry out public works projects or not, often rests in the realm of politics and political sway (Hird 1991), giving rise to the concept of "pork projects", a term used to identify pet projects of legislators that may or may not appear economically rational within the larger context of government spending. While Hird and Peresky's critique of cost-benefit analyses rests on an economic critique that reaffirms the need for economic rationality, mine here is different. I suggest that if cost-benefit analyses can be manipulated to be political justifications rather than economic rationales, then it further disenfranchises those at the political margins. Thus, when cost-benefit analysis on beach nourishment are done "correctly" they can protect expensive homes and tourist locations and disenfranchise the poor. When they are done for political reasons, they are also likely to not support the politically disenfranchised. In either case, more vulnerable communities remain unprotected. The point here is that when we talk about sea level rise and migration, we talk about whose homeland is protected by the state and whose homeland is sacrificed. If protection is not arbitrary, but rests along clear lines of class and/or those who have political power—then we can expect climate change migrants to continue to be the most vulnerable members of society.

Living in the Sacrifice Zone

The island of Isle de Jean Charles in Louisiana is experiencing some of the most dramatic consequences of relative sea level rise in the world. By 2015, the island had shrunk from its once documented 22,400 acres to an estimated 320 acres (Maldonado and Peterson forthcoming) due to a combined assault of sea level rise, erosion, storm surge inundation, and the impacts of oil dredging (a process of removing sediment in order to access oil deposits and to allow machinery and boats into shallow areas) (Maldonado 2014). The island has become a poster-child for climate change outcomes and has been featured in the New York Times under the headline, "Resettling the First Climate Change Refugees" (Davenport 2016).

While it is certain that climate change is affecting the island and the Isle de Jean Charles tribal community, there are also socio-political issues that underlie the experiences of this community. What we see playing out in this community is a clear illustration of the cost-benefit analysis system at work. In 1998 the US Army Corps of Engineers, local levee districts and Louisiana state agencies drew boundaries for the Morganza to the Gulf Hurricane Protection Project (Maldonado 2014, Fiske and Marino in press). The Morganza Project is a 70–75 mile levee, which is scheduled to be completed by 2020, and whose purpose is to protect 110,000 residents in two Louisiana parishes from the extreme impacts of hurricanes, other large storms, and erosion (Katz 2003).

The island of Isle de Jean Charles was left out of the levee project because the cost of building the levee around the island was considered too high in relationship to the benefits. In other words, Isle de Jean Charles came out on the wrong side of the cost-benefit analysis. What is really important to note is that, for years, while the community was considered too expensive to protect—there was no clearly-defined alternative plan for residents living on the island.

If cost-benefit analyses regarding infrastructure projects can be a function of the political power, as suggested by Hird (1991), then it is important to ask who lives on the island that was political sanctioned to be "sacrificed"? Isle de Jean Charles, it turns out, is home to the tribal community of the same name—a band of Biloxi-Chitimacha-Choctaw peoples who have a long history on the island, and a long history of being excluded, marginalized, and oppressed by the state (Maldonado 2014). The Isle de Jean Charles band was forcibly displaced as European and early white American settlers moved into the southeastern United States. Fleeing persecution, this group found refuge on the gulf island and made a home and sustained an economy based on fishing and catching oysters and other sea life (Maldonado 2014, Maldonado et al. 2013).

Maldonado claims that Isle de Jean Charles has become a sacrifice zone (Maldonado 2014), a term identified by Buckley and Adam and defined as "a place where human lives are valued less than the natural resources that can be extracted from the region" (2011: 171). The cost-benefit analysis of the Morganza project failed to account for this history, for the value of traditional lifestyles, and the social costs of habitually marginalizing a specific Indigenous group. Sacrifice zones are the product of decisions that have both economic and socio-political drivers. In other words, there are places and peoples who are abandoned by the state when faced with risk—contrasted with those who are not. In trade-off scenarios this is, of course, obviously the case. The interesting question here is to make explicit which communities are abandoned and whether or not these decisions are patterned in ways that consistently "sacrifice" similar groups of people in consistent ways.

In 2016, after years of dedicated work by leaders of the Isle de Jean Charles tribe, in partnership with the Lowlander Center and subject matter experts, the state of Louisiana won a competitive resilience grant from a joint venture between the federal government and the Rockefeller foundation, which included money to relocate the tribal community to a safe place off the island. This relocation has yet to occur—but here we see that the decision to protect communities is, at least, on the table.

Conclusion

The reality of climate change should and does call the global citizenry to take stock of our actions. Included in the suite of possible harms that could befall human communities as a result of climate change are impacts that push the migration of people away from their homes and homelands. Understanding the impacts of climate change outcomes on human migration, however, is a complex and challenging subject. The apocalyptic narratives that directly link a climate outcome to millions of migrants, what Bettini describes sarcastically as "climate barbarians at the gate" (Bettini 2013), do little to describe the actual series of events that push or pull migration in the lives of people on the ground. Most nefariously we can point out that these apocalyptic narratives are likely the result of xenophobic fears of immigration that exist independently of climate change at all.

What this chapter has tried to show is that climate change impacts enter into dramatically complex socio-political realities, and that those realities are rife with political obfuscation, social inequity, decision-making by powerful actors, and histories that render certain communities worthy of being saved and others suitable for sacrifice. In Alaska we see that histories of colonization affect who is in harm's way and who has access to appropriate adaptation strategies. On the East coast we see that beach nourishment is saved for tourist communities and other economically valued locations. In Louisiana we see that a cost-benefit analysis can immediately designate your community as safe or "sacrificed"—but in Louisiana we also have demonstrable proof that alternative choices—those that promote social justice—may be made.

In the end it seems that climate change risks are, to a great extent, social constructions—that, as a society we may choose whether the cost of protecting communities is something we will collectively, politically, and economically support or not. This research demonstrates some of the mechanisms that create vulnerability to climate change. Basic economic assumptions about what is valuable, for example, or historical marginalization of particular communities, are not arbitrarily co-occurring with places experiencing climate change outcomes. Instead, these social conditions are often the drivers of climate change impacts, and the drivers of climate change migration. In this realization is great hope—that we have a choice of whether to protect communities or not, and that there is no economic rationale to dissuade us from seeking out a more just future.

References

Adamo SB, de Sherbinin A (2011) The impact of climate change on the spatial distribution of populations and migration. Population distribution, urbanization, internal migration and development: An international perspective, 161

Baird R, Migiro K, Nutt D, Kwatra A, Wilson S, Melby J, Bongaarts J (1992) Human tide: the real migration crisis. People Count 2(6):1–4

Bettini G (2013) Climate barbarians at the gate? A critique of apocalyptic narratives on 'climate refugees'. Geoforum, 45, 63–72

Bronen R (2008) Alaskan communities' rights and resilience. Forced Migration Review, 31:30–32

Bronen R (2010) Forced migration of Alaskan indigenous communities due to climate change. In Environment, forced migration and social vulnerability. Springer, Berlin, Heidelberg, pp. 87–98

Bronen R (2011) Climate-induced community relocations: creating an adaptive governance framework based in human rights doctrine. NYU Rev. L. & Soc. Change, 35:357

Bronen R, Chapin FS (2013) Adaptive governance and institutional strategies for climate-induced community relocations in Alaska. Proceedings of the National Academy of Sciences 110 (23):9320-9325

Campbell-Lendrum D, Corvalán C (2007) Climate change and developing-country cities: implications for environmental health and equity. J Urban Health 84(1):109–117

Church JA, Clark PU, Cazenave A, Gregory JM, Jevrejeva S, Levermann A, Merrield MA, Milne GA, Nerem RS, Nunn PD, Payne AJ, Pfeffer WT, Stammer D, Unnikrishnan A (2013) Sea level change. In: Stocker TF, Qin D, Plattner G-K, Tignor M, Allen SK, Boschung J, Nauels A, Xia Y, Bex V, Midgley PM (eds. Climate change 2013: the physical science basis. contribution of working Group I to the fifth assessment report of the intergovernmental panel on climate Change. Cambridge University Press, Cambridge, United Kingdom and New York, NY, USA

Cooper JAG, McKenna J (2008) Social justice in coastal erosion management: The temporal and spatial dimensions. Geoforum 39(1):294–306

Davenport C, Robertson C (2016) Resettling the first American 'climate refugees'. New York Times

DeConto RM, Pollard D (2016) Contribution of Antarctica to past and future sea-level rise. Nature 531(7596):591–597

Ducker JH (1996) Out of harm's way: relocating northwest Alaska Eskimos, 1907–1917. Am Indian Culture Res J 20(1):43–71

Farbotko C, Lazrus H (2012) The first climate refugees? Contesting global narratives of climate change in Tuvalu. Glob Environ Change 22(2):382–390

Gemenne F (2011) Why the numbers don't add up: a review of estimates and predictions of people displaced by environmental changes. Glob Environ Change 21:S41–S49

Gopalakrishnan S, Smith MD, Slott JM, Murray AB (2011) The value of disappearing beaches: a hedonic pricing model with endogenous beach width. J Environ Econo Manage 61(3):297–310

Government Accounting Office (2009). Alaska native villages:limited progress has been made on relocating villages threatened byflooding and erosion. Government Accountability Office Report GAO-09-551

Hartmann B (2010) Rethinking climate refugees and climate conflict: rhetoric, reality and the politics of policy discourse. J Int Dev 22(2):233–246

Hinkel J, Lincke D, Vafeidis AT, Perrette M, Nicholls RJ, Tol RS, Levermann L (2014) Coastal flood damage and adaptation costs under 21st century sea-level rise. Proc Natl Acad Sci 111 (9):3292–3297

Hird JA (1991) The political economy of pork: project selection at the U.S. Army Corps of Engineers. American Political Science Review 85(2):429–456.

Houston JR (1996) International tourism and US beaches (No. CERC-96-2). Coastal Engineering Research Center Vicksburg, MS

Hoffman SM, Oliver-Smith A (1999) Anthropology and the angry earth: An overview. The angry earth: Disaster in anthropological perspective. Psychology Press, Rutledge, pp 1–16

Knight S (2009) The human Tsunami. The Financial, Times, p 19

Kattsov VM, Källén E, Cattle HP, Christensen J, Drange H, Hanssen-Bauer I, Vavulin S (2005) Future climate change: modeling and scenarios for the Arctic. Arctic Climate, Impact Assessment, pp 100–150

Katz M (2003) Staying afloat: how federal recognition as a Native American tribe will save the residents of Isle de Jean Charles. Louisiana. Loy. J. Pub. Int. L. 4:1

Larsen JN, Anisimov OA, Constable A, Hollowed AB, Maynard N, Prestrud P, Prowse TD, Stone JMR (2014) Polar regions. In: Barros VR, Field CB. Dokken DJ, Mastrandrea MD, Mach KJ, Bilir TE, Chatterjee M, Ebi KL, Estrada YO, Genova RC, Girma B, Kissel ES, Levy AN, MacCracken S, Mastrandrea PR, White LL (eds) Climate change 2014: impacts, adaptation, and vulnerability. Part B: regional aspects. contribution of working Group II to the fifth assessment report of the intergovernmental panel on climate change. Cambridge University Press, Cambridge, United Kingdom and New York, NY, USA, pp. 1567–1612

Maldonado J, Peterson K (Forthcoming) A community-based model for resettlement: lessons from coastal louisiana. In: McLeman R, Gemenne F (eds) The Routledge handbook of environmental displacement and migration. New York, Routledge

Maldonado JK, Shearer C, Bronen R, Peterson K, Lazrus H (2013) The impact of climate change on tribal communities in the US: displacement, relocation, and human rights. Climatic Change, 120(3):601–614

Maldonado JK (2014) A multiple knowledge approach for adaptation to environmental change: lessons learned from coastal Louisiana's tribal communities. Journal of Political Ecology 21 (1):61–82

Marino E (2012) The long history of environmental migration: Assessing vulnerability construction and obstacles to successful relocation in Shishmaref Alaska. Global Environ Change 22(2):374–381

Marino E (2015) Fierce Climate, Sacred Ground: An Ethnography of Climate Change in Shishmaref. University of Alaska Press, Alaska

Marino E, Lazrus H (2015) Migration or Forced Displacement? The Complex Choices of Climate Change and Disaster Migrants in Shishmaref, Alaska and Nanumea. Tuvalu. Human Organization 2015:341–350

Marino E, Ribot J (2012) Special issue introduction: adding insult to injury: climate change and the inequities of climate intervention

Myers N (1993) Environmental refugees in a globally warmed world. Bioscience 43(11):752–761

Neumann B, Vafeidis AT, Zimmermann J, Nicholls RJ (2015) Future coastal population growth and exposure to sea-level rise and coastal flooding-a global assessment. PLoS One 10(3): e0118571

Nicholls RJ, Cazenave A (2010) Sea-level rise and its impact on coastal zones. Science 328 (5985):1517–1520

Nosengo N (2003) Venice floods: save our city! Nature 424(6949):608–609

Oli B (2008) Migration and climate change. IOM migration research series no. 31. International Organization for Migration, Geneva, 1–68

Oliver-Smith A (ed) (2009) Development & dispossession: the crisis of forced displacement and resettlement. School for Advanced Research Press

Persky J (2001) Cost-benefit analysis and the classical creed. Journal of Economic Perspectives 15 (4):199–208

Pfeffer WT, Harper JT, O'Neel S (2008) Kinematic constraints on glacier contributions to 21st-century sea-level rise. Science 321(5894):1340–1343

Purser RE, Park C, Montuori A (1995) Limits to anthropocentrism: toward an ecocentric organization paradigm? Acad Manag Rev 20(4):1053–1089

Rahmstorf S (2007) A semi-empirical approach to projecting future sea-level rise. Science 315 (5810):368–370

Rignot E, Velicogna I, van den Broeke MR, Monaghan A, Lenaerts JT (2011) Acceleration of the contribution of the Greenland and Antarctic ice sheets to sea level rise. Geophys Res Lett 38 (5):00001

Screen JA, Simmonds I (2010) The central role of diminishing sea ice in recent Arctic temperature amplification. Nature 464(7293):1334–1337

Serreze MC, Barry RG (2011) Processes and impacts of Arctic amplification: A research synthesis. Global Planet Change 77(1):85–96

Shearer C (2011) Kivalina: a climate change story. Haymarket Books

Smith MD, Slott JM, McNamara D, Murray AB (2009) Beach nourishment as a dynamic capital accumulation problem. J Environ Economics Manage 58(1):58–71

Thompson C, Johnson T, Hanes S (2016) Vulnerability of fishing communities undergoing gentrification. Journal of rural studies 45:165–174

US Army Corps of Engineers (2007) Shoreline protection assessment: Beach nourishment, how beach nourishment works. Retrieved from https://www.horrycounty.org/Portals/0/docs/beachrenourishment/HowBeachNourishmentWorks%20(002).pdf

Warner K (2010) Global environmental change and migration: Governance challenges. Glob Environ Change 20(3):402–413

Watkins D (2015) What China has been building in the South China Sea. New York Times, 27

Chapter 12
Recovery After Disasters: How Adaptation to Climate Change Will Occur

Robert B. Olshansky

Abstract Adaptation to climate change in general, and sea level rise in particular, will be a complex process involving difficult decisions for communities. Scores of coastal cities will need to make some adjustments to rising sea level. In most cases, communities will confront disruptive new sea levels through large coastal storms and storm surges rather than as a result of slowly rising waters. Thus, adaptation to sea level rise will occur, to a great extent, through the process of long-term post-disaster recovery following these episodic disasters. If severe coastal storms are the carrier of sea level rise, then post-disaster recovery is the means of adaptation. This paper briefly summarizes what we know about the process of post-disaster recovery, with particular attention to the process of community relocation after disasters. We know that recovery is a fast-paced process with many actors, and that smart recovery requires intention, resources, and organizations designed to operate effectively in post-disaster compressed time environments. Successful recovery requires citizen involvement, and relocation in particular requires citizens to be empowered to be partners in the decisions. Still, relocation is inherently challenging, because it is expensive, residents have strong attachments to place, and relocations often disrupt social and economic networks and impede livelihoods.

Introduction

Sea level rise will be an ongoing reality in the coming decades, regardless of international efforts at greenhouse gas mitigation. Even if the nations of the world can succeed in arresting the growth in greenhouse gas emissions, the resultant changes in the climate could take hundred or thousands of years to reverse (Solomon et al. 2008; IPCC 2014). Given that sea level will be rising, property

---

R. B. Olshansky (✉)
Department of Urban and Regional Planning, University of Illinois at Urbana-Champaign, Champaign, USA
e-mail: robo@illinois.edu

owners, organizations, institutions, and local and national governments need to start developing adaptation strategies

Adaptation to climate change in general, and sea level rise in particular, will be a complex process involving difficult decisions for communities. It will be triggered by environmental changes that will impede the ability of many households to inhabit certain locations, pursue livelihoods, and/or easily access the necessities of life. In many places, the changes will be so severe so as to require significant reconstruction, localized relocation, or relocation to distant places. This process raises many social, political, economic, financial, and legal questions: At what point will environmental changes become large enough to warrant major actions such as relocation? Who makes the decision, and by what process? Would the adaptation require changes in livelihoods? Who would pay for the actions, and by what means of financing? How would community members convert their property rights in the current setting to property rights in the adapted community? And the most important question is: how can adaptation be accomplished in a manner that is equitable to all participants?

Of all climate change phenomena, sea level rise is the most clearly understood: we know that sea level will increase, we know where it will occur, and we know the approximate rate over time (Mengel et al. 2016). Scores of major world coastal cities are at risk (Hallegatte et al. 2013), and it is likely that most of them will need to make some adjustments to rising sea level. How will such adjustments occur? The rise in sea level is gradual and inexorable, generally estimated at about one-half to two centimeters per year (e.g., Church et al. 2013; Mengel et al. 2016). In most cases, however, communities will confront disruptive new sea levels through large coastal storms and storm surges rather than as a result of slowly rising waters. Not only will these episodic disasters draw attention to the problem of rising seas, the damage caused will also provide an opportunity to build back differently.

Adaptation to sea level rise will therefore occur, to a great extent, through the process of long-term post-disaster recovery. If severe coastal storms are the carrier of sea level rise, then post-disaster recovery is the means of adaptation. To understand how the process of adaptation will occur, we can draw lessons from what we know regarding post-disaster recovery processes from the recent past, especially following coastal storms. Similar issues will arise in floodplains that may experience more frequent flooding in a changed climate.

This paper briefly describes our current understanding of the process of post-disaster recovery, introduces the idea of relocation following coastal storms, summarizes the key issues and decision points for relocating communities, and then provides some illustrative examples of community relocations following disasters. The purpose of this paper is to bring to the reader's attention the existence of a large number of historic cases of community relocation after disasters and to identify some of the lessons that can be applied to future adaptation to sea level rise.

Post-disaster Recovery

Following is a brief summary of comparative research of post-disaster recovery cases over the past two decades by the author and various collaborators, primarily in the U.S., Japan, China, India, Indonesia, New Zealand, and Haiti (Johnson and Olshansky 2017; Johnson and Olshansky 2016; Olshansky and Johnson 2010; Iuchi et al. 2015; Iuchi 2014; Chandrasekhar et al. 2014; Johnson and Mamula-Seadon 2014; Balachandran 2010; Thiruppugazh and Kumar 2010; Olshansky and Etienne 2011).

Recovery is the long-term process of permanent reconstruction, extending for months and years after a disaster. It is distinguished from the process of immediate post-disaster response, which comprises emergency activities in the hours, days, and weeks following a disaster event. Recovery is a community-building process, in which a newly constructed, permanent community rises from the rubble of disaster.

It is not easy to define what constitutes a successfully completed "recovery." Speed is an important quality, but not the only one. Reconstruction after disasters offers opportunities for betterment and for fixing long-standing problems in infrastructure, land use, construction quality, the economy, and governance. Importantly, reconstruction can provide the chance to reduce the effects of future disasters by avoiding hazardous locations and improving construction quality. When is recovery complete? Recovery happens, say Alesch et al. (2009), when the community is once again a functioning system, though probably different from the original system. Furthermore, segments of the community will recover at different rates; some may not recover at all.

We know that post-disaster recovery is a complex process that involves many individuals and organizations—business, governmental, nonprofit—making decisions (Haas et al. 1977; Alesch et al. 2009). The process of city building in normal times involves many different actors of various types, and the same is true following disasters. The recovery process involves residents rebuilding, businesses resuming their operations, public agencies and utilities reconstructing and restoring services, and non-governmental organizations of all types meeting a variety of needs. Just as in normal times, individuals and businesses make development decisions within a context set by governmental actions and rules, the economy, and the actions of other individuals and organizations. In most societies, no single entity is in charge of the reconstruction, just as no single entity controls the normal process of urban development.

When a community is devastated by disaster, it needs large amounts of money to pay for the labor and materials for rebuilding—essentially the same as it cost to build the community in the first place. Regardless of whether it is a large developed city or an urbanized region in a developing nation, costs of such magnitude are overwhelming to the affected area. A central aspect of recovery is a search for sources of financing to accomplish this great task. Manifestations of this search for funding include: reconstruction plans, lobbyists, political favors, conflicts between governmental factions, claims with insurance companies, deals with developers, and negotiations with government agencies such as FEMA; the international arena also includes bilateral aid, multilateral aid organizations, international NGOs, and all the restrictions accompanying funding by these entities.

With so many actors involved in recovery, the other key ingredient is information. Each actor can operate more effectively and efficiently to the extent that they know what the other actors—neighbors, utilities, governments, funding sources—are doing and intend to do. Broad information flows, efficient use of communication networks, and design of information clearinghouses and coordinating bodies are all ways of fueling and lubricating the recovery process.

Above all, the characteristic that distinguishes post-disaster conditions from normal times is what we call "time compression"—a compression of activities in time and focused in space (Olshansky et al. 2012). A disaster results in a sudden loss of capital stock—which is normally replaced gradually over time—and an immediate need for its replacement.

Time compression is what makes post-disaster recovery a unique setting for city-building processes. It means that actions, financial flows, and information all occur in a very short time period, often simultaneously. This has several implications, leading to a variety of phenomena that make post-disaster recovery especially challenging and confusing for participants. Perhaps the most significant challenge is that actions often occur faster than information can support and coordinate them; each actor simply cannot wait to receive and absorb relevant information about the actions of others. In this way, recovery actors feel that they are in the "fog of war," proceeding blindly toward an unknown goal in a game with unknown rules.

Some phenomena compress in time more easily than do others. For example, commerce can restart quickly in temporary settings, but permanent construction takes more time. Some funding sources flow more readily than others; insurance tends to come more quickly, and the various pots of governmental funding all have their own timelines. The result is that many actions seem to be "out of sync," and, to recovery actors, it feels like they are in a very different world than in normal times.

Time compression of losses creates sudden opportunities for change: reduction of risk from future disasters, replacement of obsolete infrastructure, improvements in transportation and urban design, new industries, reinvention of urban services, and more equitable housing opportunities. Positive change, however, requires deliberation, which is difficult to do in compressed time.

A dilemma for many organizations in post-disaster recovery is the tension between speed and deliberation: between quickly replacing what was lost or taking time to think about how to build back better. In fact, under time compression, speed is difficult to avoid, as most actors feel compelled to rebuild as quickly and expeditiously as possible.

The challenge, then, is to endeavor to deliberate faster, so as to enhance the chances for long-term betterment. One way to do this is to increase the bandwidth of communications, by generating and distributing more information or creating new channels of communication that would be redundant in normal times. This could be done by means of a data clearinghouse, a data coordination office, and/or a well-staffed recovery communications office. Another approach to increasing information flows is to create a council of critical agencies or stakeholder organizations. For example, following Hurricane Sandy, which affected five states, the federal Hurricane Sandy Rebuilding Task Force consisted of 23 task force members consisting primarily of cabinet-level officers, with an advisory group consisting of state, local, and tribal officials from the affected areas (Hurricane Sandy Rebuilding Task Force 2013).

Other ways to act strategically include: increase planning capacity (add more planning staff, increase opportunities for citizen involvement), decentralize (devolving reconstruction decisions to more local levels), or iterate (focus on easier problems first and deal with more severely damaged areas more deliberately). An effective way to add planning capacity is to pay for local planners or facilitators to work with affected residents in helping them to plan for recovery, access recovery resources, and share information with their neighbors. Variants of this were applied in Kobe, Japan, following its 1995 earthquake; in New Orleans following 2005 Hurricane Katrina; in Maharashtra and Gujarat, India, following earthquakes in 1993 and 2001; and in Yogyakarta, Indonesia, following a 2006 earthquake (Johnson and Olshansky 2017). Probably the largest-scale example of decentralization was the central government's response to the great Wenchuan, China earthquake of 2008. The Chinese government assigned 19 wealthy eastern provinces to 24 earthquake-affected counties and cities in western China, and required them to offer assistance of at least one percent of their last budget to support earthquake reconstruction projects (Johnson and Olshansky 2017).

One of the best ways to facilitate efficient and timely post-disaster deliberation, of course, is to begin the deliberations beforehand. Plans can be carried out much more quickly if a locality has processes in place before the disaster (Schwab 2014). Pre-disaster plans can help improve both the speed and quality of post-disaster actions by having a community consensus already in place for priorities and policies after the disaster. As a result, disasters can create opportunities for communities to take actions for community improvement that may have been difficult to accomplish in normal times (Spangle Associates 1997).

The compressed time of post-disaster recovery requires different types of institutional arrangements than in normal times. Governments can create new post-disaster organizations to improve coordination among agencies. These can vary from central government agencies that can provide funding and command compliance, such as Japan's National Reconstruction Agency following the 2011 tsunami, to coordinating bodies, such as the 2013 Hurricane Sandy Rebuilding Task Force in the U.S. (Johnson and Olshansky 2017). At the local level, communities and states can form recovery councils, designed to improve communication and coordination of post-disaster plans and actions among a variety of public and private recovery organizations. In all cases, governments need to create organizations that can allocate and track the rapid flows of money and approve and monitor the large number of construction projects, because this increased workload exceeds their normal processing capacity. Examples include: Indonesia's Rehabilitation and Reconstruction Agency (*BRR—Badan Rehabilitasi dan Rekonstruksi*) for Aceh and Nias, created after the 2004 Indian Ocean Tsunami; and the Gujarat State Disaster

Management Agency, created after a 2001 earthquake in Gujarat, India. Even with such mechanisms, however, governmental bureaucracies are inherently poorly suited for recovery; they are not designed to act quickly, and they lack capacity to innovate. New local non-governmental and community-based organizations always emerge following disasters, to fill the gaps in services that governments cannot effectively provide. In fact, such organizations are the keys to successful recovery. Newly emergent locally-based NGOs are another means of increasing information channels after disasters.

In summary, recovery is a complex reality of many actors seeking resources (money, technical assistance), starved for information, and constrained by time. This dynamic (actors, resources, information, time constraint) has implications for institutional design and budgeting. The key lesson is that smart recovery requires intention, resources, and organizations designed to operate effectively in post-disaster compressed time environments.

Recovery After Coastal Storms

The most feasible way to successfully adapt to sea level rise is to relocate to higher ground. Coastal storms typically provide both the harbinger of rising sea levels as well as the opportunity, via the rebuilding process, to relocate. But the process is complicated, and successful relocation requires years of preparation before disaster strikes.

Relocation adds additional complexity to recovery and reconstruction processes. In addition to practicing smart recovery management—intentionally applying financial and information resources, and creating and supporting appropriate organizations—those seeking to achieve post-disaster relocation need to perform detailed planning, acquire additional resources, and implement a complex set of actions to move communities from one place to another. Relocations, at various scales, are relatively common features of reconstruction after great disasters, which include not only coastal storms, but also earthquakes, tsunami, volcanic eruptions, and landsliding. Truly successful post-disaster relocations are rare, however, for a variety of reasons:

- They are expensive, because of their administrative costs as well as both land and construction/demolition costs at both the old site and the new one.
- They require available land.
- They involve legal issues surrounding property rights.
- They require considerable planning, which takes time and resources.
- They are politically contentious.
- It is difficult to satisfy all stakeholders' concerns regarding equity.

Perhaps most importantly, residents resist relocation because they have reasons for living where they do. These include practical reasons of livelihoods, amenities, affordability, and social networks, as well as intangibles such as attachment to place. People resist moving, even if their homes have been flooded or their fields buried in volcanic ash. For all these reasons, except in the rare cases where communities are completely buried by landslides or water, experience tells us that to be successful, relocation must be generally voluntary.

Steps in Post-disaster Relocation

Post-disaster relocation involves numerous steps and a broad range of key decisions by all the stakeholders. All relocations involve some form of purchase of the property or property rights in the hazardous area. Relocation programs differ by the degree of involvement in resettlement: some provide a new site, whereas others simply purchase the hazardous property. For relocations that provide a new site, some intentionally move the entire community as a whole, whereas others simply offer sites for individual choice. Some of they key decision steps include.

The Risk Decision

- Is the move based on an event that makes it difficult to occupy the site, or is it based on the likelihood of reoccurrence of the disaster in the future?
- Does the hazard pose a threat to lives, or livelihoods, or property (or combinations of the three)?
- Who made the risk determination, and were residents consulted?
- Are there alternatives to complete community relocation, such as rearranging properties on the site or hazard mitigation actions?

Social and Livelihood Issues

- Why do people currently live in this location, and to what extent do their livelihoods depend on it?
- Would a distant move disrupt their access to their livelihoods?
- How might the relocation disrupt social ties?

The Process of Deciding to Leave One Place

- By what political and social processes do the residents and community organizations decide to relocate?
- Is it collective relocation to a new place, or individual relocation to many new places? Does it involve all the properties in the community, or just some of them?

The Process of Moving to Another Place

- What is the process for selecting the new location, according to what criteria?
- To what extent are residents involved in selecting the new site?
- What is the process of site planning at the new location, and to what extent are residents involved?
- How are properties allocated at the new location to owners and tenants?

Property Rights, Sources of Funding, Financing

- What property rights do residents give up at the old location, and what are their property rights at the new location? By what financing mechanisms do people give up value in the old place and gain it at the new one?
- Do tenants have rights?
- What are the sources of funds and of financing for land and for construction?
- What organization manages this process?
- Who constructs the new buildings? Property owners? Multiple contractors hired by property owners? Mass construction contractors hired by the government or construction manager? Nonprofits?

It also is important to appreciate that relocations are always connected to other pre-existing issues. Relocations might provide regional benefits, if they support other planning goals, such as regional transportation, environmental protection, or housing affordability. Conversely, they are often tangled up in local politics, in which the relocated community may be a pawn in larger power struggles that predate the disaster; sometimes, for example, local governments or the business elite may desire to move low-income or indigenous communities from certain locations. Finally, it is always important to consider who might benefit financially from the relocation, such as from land sales, new transportation access, or construction contracts.

Examples of Post-disaster Relocation

Some of the challenges of post-disaster relocation can be illustrated by means of examples from several recent disaster recovery cases. The 2004 Indian Ocean tsunami devastated coastal areas in 16 countries (Synolakis and Kong 2006). In India, it affected 2260 km of coastline, primarily in the state of Tamil Nadu (Murty et al. 2006). Maximum tsunami runup heights ranged up to a maximum of 12 m in Nagapattinam, where it also inundated areas up to three kilometers inland, as it washed away houses, damaged harbors and bridges, and inundated farmland (Maheshwari et al. 2006). The Government of India chose to reconstruct at levels higher than those affected by the tsunami and arranged for NGOs to perform the housing reconstruction (Chandrasekhar 2010). By 2009, over 18,000 units of new housing had been constructed in inland locations. For those owners who were required to relocate, however, few of them had actually moved out of their old homes. Reasons for this included: need for fishing households to stay close to the sea, culturally inappropriate housing design, poor construction quality, incomplete infrastructure in new location, and fear that once they vacated the coastal location the state would encourage new resort developments that would limit their access to the sea. The most significant reason involved their inability to easily continue their fishing livelihoods if they moved inland. In addition, the culturally inappropriate housing designs reflected an insufficient citizen involvement process in the planning phase.

The 2004 Chuetsu Earthquake in Japan affected a rural mountainous area, causing numerous landslides and road failures (Iuchi 2014). The aging, depopulating settlements in the area were faced with decisions to rebuild or relocate. One municipality offered government reconstruction programs with subsidies to encourage rebuilding in place, whereas the adjacent municipality offered programs that paid for relocation to more urbanized areas in the flatlands. It turned out that the government incentives to relocate or rebuild were not enough to over ride livelihood-based decisions. The primary factor in households' resettlement decisions (rebuild, relocate as a village, or relocate individually) was access to livelihoods: whether they preferred to continue farming in the mountains, or whether households preferred access to jobs and services in the urban areas.

Following Hurricane Katrina in 2005, the State of Louisiana, through its "Road Home" program, offered over $8.6 billion of Federal block grant funds from the Department of Housing and Urban Development (HUD) to homeowners, giving them the choice to either rebuild in place or sell their property to the state (Green and Olshansky 2012). Given the disastrous flooding of the entire city, the fact that much of the city sits below sea level, and the risk of future storms, this program provided households with the opportunity to sell their damaged properties and move to higher ground in the New Orleans area or to leave the region altogether. Of the nearly 45,000 New Orleans households that participated in Road Home, over 89% of them elected to rebuild (although not all have been financially able to do so). Even in the most deeply flooded neighborhoods (mean depth 6.2–7.5 feet), 52–70% decided to rebuild. Given the financial opportunity to leave, attachment to place kept most households exactly where they were when Katrina struck. New Orleans has deep traditions stretching back for generations, and most New Orleanians are strongly attached to the city and often to their neighborhoods as well. As a result, most residents were committed to rebuilding their neighborhoods.

The Great East Japan Earthquake and Tsunami of 2011 severely affected approximately 300 km of coastline, inundated 561 km^2 of land, and temporarily displaced up to 470,000 people (Iuchi et al. 2015). Supported by a new National Reconstruction Agency and a national reconstruction commitment of 25 trillion yen (approximately $250 billion in US dollars), four prefectures and 81 local governments developed reconstruction plans that involved a combination of seawalls, land raising, and residential relocation to higher areas. This case illustrates the challenging mechanics of implementing a massive relocation effort, even one with consistent rules and strong national government financial and technical support. As of this writing, it has been nearly 5½ years since the tsunami, and although some of the land preparation and new housing are complete, most is not. Some of the time-consuming factors include: decisions regarding whether to elevate in place or relocate, identification and acquisition of lands for relocation sites, bureaucratic requirements involved with so many construction projects, lack of sufficient local government staff to manage so many projects, coordination between different aspects of reconstruction, and some local controversies regarding seawall heights and locations.

Probably the most notable relocation success in the US would be the residential relocations following the Grand Forks, North Dakota flood of 1997. The flood, in April 1997, damaged 83% of the city's houses, displaced 90% of the 52,500 residents, and the flood and accompanying fire devastated the city's downtown buildings (HUD Exchange 2008). Using federal recovery funds, the city purchased more than 800 residential and commercial properties along the river and turned it into a greenway park. While most of the residents were able to use their buyout proceeds to relocate, the city also added to the housing supply by contracting with a nonprofit builder to construct 180 new homes in two new subdivisions on higher ground. Having a well-known and well-defined flood-prone area supported the city's decision to prohibit rebuilding, which facilitated the buyout of all the properties. The process was also successful because the city made all the decisions within the first three months, and they worked actively with each household as case managers. The existence of available land on the west side of town helped to keep many households in the city. The new subdivisions were completed and sold within approximately two years of the flood.

Although the Grand Forks case was unusual because of its scale and the construction of a housing subdivision for some of the relocated households, buyouts of small groups of floodprone properties have been common in the U.S. since the early 1990s. Although no comprehensive tabulation exists, FEMA has funded the purchase of at least 7000 floodprone properties nationwide since 2003 at a cost of over $417 million (Cater and Benincasa 2014). As a result of Hurricane Sandy, New York and New Jersey expect to purchase up to 1000 and 1300 properties, respectively (Henderson 2015). FEMA pays for buyouts of floodprone properties because, in the long run, it is less expensive than issuing insurance payments under the National Flood Insurance Program to properties that may flood repeatedly over the years. Under this program, localities, states, and FEMA work together to identify the most highly floodprone areas and then offer to buy the properties at pre-flood fair market value (FEMA 2014). To initiate a buyout project, the locality must agree to permanently designate the purchased properties as open space. These programs succeed because they strategically identify the most hazardous locations and work with neighborhoods to gain agreement; if residents don't agree, then the buyout does not proceed. In many cases, communities identify these areas ahead of time and plan for eventual relocation before the next flood. Typically in the U.S., where livelihoods are not usually tied to neighborhood location, place attachment—to a seashore, lakeshore, or riverfront—is the main reason for unsuccessful buyout attempts. Finance is also a reason in many unsuccessful cases, especially along rivers, where low market value may reflect the flood hazard: people buy these properties because they are affordable, and a fair market value purchase may give them few viable relocation options nearby.

As these few examples help to illustrate, relocations are expensive, administratively challenging, and go against residents' attachment to place, which is sometimes connected to their livelihoods. Under some circumstances, however, residents will agree to relocation after a disaster. Examples of past cases suggest relocation has a better chance of being successful if: it is voluntary, residents have been involved in the decision, they appreciate the risks they face from natural hazards, possible relocation had been discussed prior to the disaster, relocation funding is sufficient, and they have a viable place to which to relocate.

Conclusions

With regard to sea level rise, post-disaster moments provide the best opportunities for adaptation by means of relocation. Communities need to be prepared to take advantage of these moments. Success, however, requires development of plans and policies ahead of time. This means that coastal communities should be developing such policies now.

Because residents have strong attachments to place, relocation must be voluntary in order to succeed. Households need to make their own decisions, based on their access to livelihoods, financial circumstances, social networks, and sense of place. Thus, relocation programs require substantial citizen involvement, in which residents are empowered to be partners in the decisions.

Smart recovery planning processes, whether or not relocation is included, involve considerable information resources (including data, legal resources, and technical assistance) and active citizen involvement. Such processes requires intention and the commitment of resources. Paying for the rebuilding of structures is only one of the costs of recovery; successful recovery also involves mobilizing human and social capital, as well as ensuring fairness and transparency. Communities seeking to pursue successful post-disaster relocation need to be prepared to provide resources for information and citizen involvement.

Finally, even if communities design effective processes and provide appropriate resources, they need to realize that relocation is still inherently challenging. In particular, identifying appropriate relocation sites near coastal areas will be difficult, and financing coastal relocations will be a challenge that national governments will need to face.

References

Alesch DJ, Arendt LA, Holly JN (2009) Managing for long-term community recovery in the aftermath of disaster. Public Entity Risk Institute, Fairfax, VA

Balachandran B (2010) Planning the reconstruction of Bhuj. In: Patel S, Revi A (eds) Recovering from earthquakes: response, reconstruction, and impact mitigation in India. Routledge, New Delhi

Cater F, Robert B (2014) Map: FEMA is buying out flood-prone homes, but not where you might expect. NPR Cities Project, 20 Oct 2014. http://www.npr.org/sections/thetwo-way/2014/10/20/357611987/map-femas-buying-out-flood-prone-homes-but-not-where-you-might-expect

Chandrasekhar D (2010) Understanding stakeholder participation in post-disaster recovery (case study: Nagapattinam, India). Doctoral Dissertation, University of Illinois at Urbana-Champaign

Chandrasekhar D, Zhang Y, Xiao Y (2014) Nontraditional participation in disaster recovery planning: cases from China, India, and the United States. J Am Plann Assoc 80(4):373–384

Church JA, Clark PU, Cazenave A, Gregory JM, Jevrejeva S, Levermann A, Merrifield MA, Milne GA, Nerem RS, Nunn PD, Payne AJ, Pfeffer WT, Stammer D, Unnikrishnan AS (2013) Sea level change. In: Stocker TF, Qin D, Plattner GK, Tignor M, Allen SK, Boschung J, Nauels A, Xia Y, Bex V, Midgley PM (eds) Climate change 2013: the physical science basis. Contribution of working group I to the fifth assessment report of the intergovernmental panel on climate change. Cambridge University Press, Cambridge, United Kingdom and New York, NY, USA

HUD Exchange (2008) Recovery snapshot: grand forks residential buyout program, CDBG-DR Training, 23–24 Apr 2008. https://www.hudexchange.info/training-events/courses/cdbg-disaster-recovery-compliance-training/

FEMA (2014) For communities plagued by repeated flooding, property acquisition may be the answer. Press release SRFO-NJ NR-023, 28 May 2014. https://www.fema.gov/news-release/2014/05/28/communities-plagued-repeated-flooding-property-acquisition-may-be-answer

Green TF, Olshansky RB (2012) Rebuilding housing in New Orleans: the road home program after the Hurricane Katrina disaster. Housing Policy Debate 22(1):75–99

Haas JE, Kates R, Bowden M (eds) (1977) Reconstruction following disaster. MIT Press, Cambridge, MA

Hallegatte S, Colin G, Robert JN, Jan C-M (2013, Sep) Nature climate change, vol 3. https://doi.org/10.1038/nclimate1979

Henderson T (2015) Rising tide of state buyouts fights flooding. Stateline, Pew Charitable Trusts, 3 Dec 2015. http://www.pewtrusts.org/en/research-and-analysis/blogs/stateline/2015/12/03/rising-tide-of-state-buyouts-fights-flooding

Hurricane Sandy Rebuilding Task Force (2013) Hurricane Sandy rebuilding strategy: stronger communities, a resilient region. U.S. Department of Housing and Urban Development. https://portal.hud.gov/hudportal/documents/huddoc?id=hsrebuildingstrategy.pdf. Retrieved 21 Aug 2013

International Panel on Climate Change (IPCC) (2014) Climate change 2014: synthesis report. In: Core Writing Team, Pachauri RK, Meyer LA (eds) Contribution of working groups I, II and III to the fifth assessment report of the intergovernmental panel on climate change. IPCC, Geneva, Switzerland, 151 p

Iuchi K (2014) Planning resettlement after disasters. J Am Plann Assoc 80(4):413–425

Iuchi K, Elizabeth M, Laurie J (2015) Three years after a megadisaster: recovery policies, programs and implementation After the Great East Japan Earthquake. In: Santiago-Fandiño V et al. (eds) Post-tsunami hazard. Springer International Publishing, pp 29–46

Johnson LA, Mamula-Seadon L (2014) Transforming governance: how national policies and organizations for managing disaster recovery evolved following the 4 september 2010 and 22 february 2011 Canterbury earthquakes. Earthquake Spectra 30(1):557–605. https://doi.org/10.1193/032513EQS078M

Johnson LA, Olshansky RB (2016) After great disasters: how six countries managed community recovery. Policy focus report. Lincoln Institute of Land Policy, Cambridge, MA

Johnson LA, Olshansky RB (2017) After great disasters: an in-depth analysis of how six countries managed community recovery. Lincoln Institute of Land Policy, Cambridge, MA

Maheshwari BK, Sharma ML, Narayan JP (2006) Geotechnical and structural damage in Tamil Nadu, India, from the December 2004 Indian Ocean tsunami. Earthquake Spectra 22(S3): 475–493

Mengel M, Levermann A, Frieler K, Robinson A, Marzeion B, Winkelmann R (2016) Future sea level rise constrained by observations and long-term commitment. Proc Natl Acad Sci 113 (10):2597–2602. https://doi.org/10.1073/pnas.1500515113

Murty CVR, Jain SK, Sheth AR, Jaiswal A, Dash SR (2006) Response and recovery in India after the december 2004 great sumatra earthquake and Indian Ocean tsunami. Earthquake Spectra 22 (S3):731–758

Olshansky RB, Etienne HF (2011) Setting the stage for long-term recovery in Haiti. Earthquake Spectra 27(S1):S463–S486. https://doi.org/10.1193/1.3633096

Olshansky RB, Johnson LA (2010) Clear as mud: planning for the rebuilding of New Orleans. American Planning Association, Chicago, IL and Washington DC

Olshansky RB, Hopkins LD, Johnson LA (2012) Disaster and recovery: processes compressed in time. Nat Hazards Rev 13(3):173–178. https://doi.org/10.1061/(ASCE)NH.1527-6996. 0000077

Schwab J (ed) (2014) Planning for post-disaster recovery: next generation. Planning advisory service report 576. American Planning Association, Chicago, Illinois

Solomon S, Plattner GK, Knutti R, Friedlingstein P (2008) Irreversible climate change due to carbon dioxide emissions. Proc Natl Acad Sci 106(6):1704–1709. https://doi.org/10.1073/pnas. 0812721106

Spangle Associates (1997) Evaluation of use of the Los Angeles recovery and reconstruction plan after the Northridge earthquake. Spangle Associates, Portola Valley, CA

Synolakis CE, Kong L (2006) Runup measurements of the december 2004 Indian Ocean tsunami. Earthquake Spectra 22(S3):S67–S91

Thiruppugazh V, Kumar S (2010) Lessons from the Gujarat experience for disaster mitigation and management. In: Patel SB, Revi A (eds) Recovering from earthquakes: response, reconstruction, and impact mitigation in India. Routledge, New Delhi, pp 223–237

Part V
Responding to Climate Change: Priorities, Perspectives, and Solutions

Chapter 13
The Climate-Change Challenge to Human-Drawn Boundaries

Eric T. Freyfogle

Abstract The plights of climate-change migrants raise serious questions about human-drawn territorial boundaries, among nation states, among political subdivisions, and indeed between and among private property owners. These questions go beyond matters of social justice and individual rights to include matters relating to the abilities of local people everywhere to live in ways that keep them, and their lands and waters, healthy. The challenges of migrants thus are usefully considered as part of a larger inquiry into how we live in nature and what it will take for us and our natural homes to thrive over time. That larger inquiry needs to pay particular attention to the root causes of our misuses of nature, including (but hardly limited to) our behaviors that stimulate climate change. Good policies would help people everywhere succeed at this foundational task of living on land without degrading it, and they would help migrants through means that respect local efforts to live rightly in nature. Wisely drawn and understood, human-drawn boundaries of all types could help make this overall goal possible, likely giving rise to political and proprietary boundaries that are, in various ways, both more and less permeable than those we have today.

The global problem of climate change offers an especially vivid reminder of how nature's processes commonly ignore the boundaries we draw on our maps and our lands, both our political boundaries, separating state from state and province from province, and our private property lines. At some level we know this, of course, that these human-drawn boundaries are physically artificial. Wildlife, rivers, exotic species, diseases, all move across them, as does pollution. Human activities routinely transcend them, with resources, commodities, manufactured goods, ideas, values, money all crossing our borders. Our moral concerns can similarly spread widely: People in one country can care greatly about people in distant places. They can worry about losses of species and ecological declines in places far from where they live.

E. T. Freyfogle (✉)
University of Illinois, 504 E. Pennsylvania Ave, Champaign, IL 61820, USA
e-mail: efreyfog@illinois.edu

© Springer International Publishing AG, part of Springer Nature 2018
C. Murphy et al. (eds.), *Climate Change and Its Impacts*,
Climate Change Management, https://doi.org/10.1007/978-3-319-77544-9_13

Among the ill effects of climate change is that the homes of many people are becoming less habitable, pressuring them to move. Some places—some coastal zones, for instance—are slowly becoming uninhabitable. Other places can no longer support the populations that they have and reductions are needed. Then there are the many places—floodplains along many rivers, for instance—that residents ought to evacuate so we can devote the lands to uses that better sustain natural processes and wild species. Some lands are perhaps best used, from a communal perspective, as corridors for wild animals and plants to respond to climate change, or they might best serve to contain and filter floodwaters heightened by climate change. In short, climate change is affecting how we might best occupy lands and waters, in inland areas as well as coastal zones. As it does so, it presses against and calls into question the boundaries we have imposed, boundaries that, to varying degrees, allocate power over land and thus over the people whose lives depend on access to that land.

Most vivid to many observers today is the plight of human climate-change migrants, both people whose homelands suffer inundation and those whose lands are so affected by worsening droughts—much of war-torn Syria, for instance—that settlement patterns if not civil peace are radically upset.[1] Their plights pose an expanding humanitarian crisis of widespread suffering. Sometimes victims are categorized in ethnic and national terms. More often they are described simply as individuals and families, buffeted by forces beyond their control. They need new homes, new places to go. They need to cross boundaries to get there. And typically, they need to end up, or would end up if allowed to do so, in places where they have no legal right to settle.

The plight of such climate migrants poses stark questions about territorial control, by private owners as well as states.[2] As many have observed, to control nature is to wield power over other people—plainly so, for instance, in desert lands where water supplies are tightly controlled. What makes such territorial controls morally legitimate, and how much legitimacy attaches to the arrangements we now have? Such questions quickly lead to others. Are the planet and its resources in moral terms the common property of all, with each person entitled to a fair share?[3] Are individual rights at stake here, perhaps some right for climate migrants to settle in new places notwithstanding the territorial claims (and objections) of existing inhabitants? In broad terms, does climate change give us cause to weaken territorial powers by making such boundaries (existing and perhaps future) more permeable?

A common response to migration pressures has been to look to the world's wealthiest nations and demand that they open their borders to more migrants. Three related rationales for this stance are often put forth: Such nations have played bigger

---

[1]The many good works on the subject include Welzer (2012), original German ed. 2008. On Syria, see Fountain (2015).

[2]Particularly good cultural and ethical assessment are Orr (2016) and Northcott (2007).

[3]See, for instance, Oberman (2017). I refer here, of course, to the relative claims among humans, putting to one side the interests of other life forms and natural communities as such.

roles in causing climate change; they are the places where migrants often want to head; and they are economically better able to absorb migrants. The first argument comes across as a claim of fairness: those who cause a problem should be responsible for it. The second claim appears more as a matter of individual rights possessed by the migrant themselves, rights that arguably deserve respect by actors capable of respecting them. The final claim in practice mixes these considerations and adds the element of practicality.

This essay takes up the pressing topic of climate-change and human-drawn boundaries. It is a topic intertwined with many others, environmental as well as social, which means it is not well handled (as it too often is) as a discrete issue. Because the topic sprawls so widely one can do little more in an essay than simply frame the topic, staking out the main questions and proposing how we might best approach them.[4] Some of the main points can be covered rather quickly and easily. Others can be raised and presented but, given space constraints, not developed with enough fullness to convince readers for whom the points seem new or questionable.

In brief and to look ahead, the plight of climate-change migrants is not best framed in terms of individual rights and, further, it is only secondarily a matter of social justice, for migrants and others. Indeed, to focus on migrants chiefly as collections of individuals is to accentuate modes of thinking that come at rather high cost—higher than we recognize—to our collective efforts to live sensibly on the planet. It is similarly diverting and costly to dwell on international borders as a distinct problem, paying attention to them without, at the same time, attending also to similar, human-drawn boundaries on lands and waters. Climate change does call upon us to reconsider all such boundaries and, in some ways, make them more permeable. Yet, particularly when linked with other environmental ills—as it needs to be–climate change also give reasons for us to make the boundaries we've created stronger than they have been, and to afford the people who inhabit a particular place greater power to resist outside forces and pressures, at least when they use that power well. Looking ahead, we may in fact need, in many places, more rather than less power residing at small spatial scales, more rather than less territorial autonomy, if we are to foster ways of living that enable people and surrounding nature to flourish over the long run.

Ultimately, human-drawn boundaries should exist when and to the extent their presence and enforcement of them help us flourish; help us live better, in nature and with one another. We are, at the moment, a good way from seeing clearly what such pleasing, health-promoting settlement patterns might best look like and a far way too from knowing what it will likely take for us to embrace them. Most of the work that lies ahead will have to do with changes in our dominant cultures. This overdue

---

[4]Many of the points made here I develop much further, in somewhat different contexts, in other writings. The broader points about humans and nature and about good land use (what it means, why it is important) are explored in *Our Oldest Task: Making Sense of Our Place in Nature* (Chicago: University of Chicago Press, 2017) and in *A Good That Transcends: How U.S. Culture Undermines Environmental Reform* (Chicago: University of Chicago Press, 2017). These works and others cited below offer citations to the relevant literatures.

cultural reform includes, with respect to our dealings with nature and our environmental ills, a marked shift away from our present preoccupation with individual rights and individual preferences. Only when and as this reform work is done well will it become feasible for us to make sound decisions about our human-created boundaries and to determine when and to what extent these boundaries deserve respect.

We can rephrase these points and present them in more logical order:

- The human ills caused by climate changes, migration pressures included, are best explored in the larger context of how we live in and with nature generally. This broad perspective differs from the common tendency to treat climate change in isolation.
- Before committing to any solutions to these ills—and, indeed, before criticizing current arrangements in much detail—it makes sense to slow down and dig deeply and expose the root causes of our ills in human behavior. Long-term solutions need to confront root causes. On this point, too, the perspective proposed here deviates from that common impulse to jump directly to solutions.
- The solutions we ultimately embrace to mitigate these ills need to solve multiple problems concurrently: It is hazardous and likely unproductive to try to help migrants in isolation. This is, in brief, a central claim of this essay. Even as we seek to help migrants we should also help people everywhere use their lands and resources in sound, enduring ways; in ways that keep nature healthy, diverse, and resilient while also sustaining human flourishing. We won't move far in that direction without addressing root causes, particularly cultural ones, which is to say the road to healthy lands will entail rather big changes in modern culture—another of this essay's central claims.
- The boundaries we draw and re-draw on the land should help promote this kind of overarching vision of nature and culture, of people everywhere dwelling sensibly in their homelands and helping sustain the good condition of larger landscapes. To a large extent, this means evaluating and reconceiving our boundaries from various spatial perspectives in terms of what they keep out and what they let in. Do they promote sound living at the local level? Do they unite people at larger spatial scales? Do they promote solutions to regional, continental, and global problems—social justice problems and environmental ills—as well as local ones? Soundly conceived boundaries can help; poorly designed boundaries can frustrate and degrade.

Respecting Nature

Four important observations can help frame the larger topic. The first two should be evident enough. The remaining two may seem less obvious.

For starters, climate change is only one of many large-scale environmental ills that transcend borders and that require, for resolution, collective responses at scales equal to the problems themselves. Large-scale animal migrations offer one example.

Animal survival often depends on seasonal migrations, on land, in water, and by air. It depends on the maintenance of physical conditions that satisfy the year-long biological needs of the migrating animals.[5] Similarly, growing dead-zones in estuaries typically result from land-uses unfolding over broad, trans-boundary catchment basins.[6] Solutions for these dead-zones require coordinated responses throughout a basin. A third illustration: Alterations of interstate river flows by upstream states can greatly harm peoples, lands, and wildlife across national borders. The many problems of interstate waters extend well beyond pollution and overall diversions of water, as the case of Bangladesh illustrates.[7] A downstream coastal nation, Bangladesh is kept alive by the silt-laden water flows of its immense rivers. The transported silt is required to sustain fertility and counterbalance natural land subsidence. The rivers' powerful water flows are similarly needed, not just for transport and irrigation, but to counteract salinity intrusion from the ocean, salinity that contaminates drinking water, kills coastal vegetation, and worsens storm damage. Sensible river management requires action at the catchment-basin level, including a careful integration of land and water uses. Beyond physical interconnections such as these, there are also, as noted, moral concerns that similarly transcend borders, such as the legitimate concerns of people in one place about biodiversity losses and ecological declines in distant lands.

These border-transcending problems—foundation point two—reach across boundaries that exist within nations as well as between them. Thus, the hypoxia problem in the Gulf of Mexico, near the outfall of the Mississippi River, arises entirely due to activities within a single nation. That nation, though, comprises many states. States upstream on the River and its tributaries often care little about problems far downstream. Given the ways sovereign power and responsibility are divided, state lines have real consequences. Within the various states are the networks of property boundaries, where the same mentality—ignoring effects that cross lines—is similarly evident. In the instance of some environmental problems (toxic emissions, for instance), the solution is to reduce if not halt the externality (as economists have termed it). In many settings, though, a different response is needed, some form of coordinated planning and action by the parties controlling the adjacent territories. Wildlife migrations fit into this category: habitat protection is needed all along a migration route.

The problems posed by these internal boundaries—private property as well as jurisdiction lines—become more manifest as we get clear on the specific steps required to deal with large-scale problems. For multiple reasons (to offer one example), levees on major rivers need breaching so that rivers and aquatic life can

---

[5]A much-cited work by leading American conservation biologists is Noss and Cooperrider (1994).
[6]Schindler and Vallentyne (2008).
[7]The plight of Bangladesh, as an illustration of the challenges facing weak, downstream states, is ably explored in Romin Tamanna, "Rivers, Ecological Health, and Justice: International Watercourses and Long-Term Legal Reform," Ph.D. diss., University of Illinois at Urbana-Champaign, 2017.

interact with their historic floodplains.[8] Typically this will entail moving private owners and their activities out of the floodplains, with or without their consent. Promoting biodiversity similarly will require large-scale, coordinated alterations in dwelling patterns, again without real consent. In broad terms, many of the steps required to improve our ways of living in nature will necessitate vast shifts in today's settlement patterns, which is to say actions that transcend and weaken existing boundaries.

The third framing element has to do with where we want to end up. In some way we need to learn to live well in and with nature, in ways that keep the land itself healthy, diverse, and beautiful while also promoting long-term human flourishing. However we capture it in words, something like this normative vision ought to be our guiding ideal, something that we embrace collectively as a shared goal, as a common good, and not merely something that emerges by summing up individual preferences.[9] Put otherwise, we should collectively be using lands and waters in good ways while avoiding the abuse of them, with "abuse" defined here as interactions with nature that appear morally or prudentially wrong when evaluated under our all-things-considered standard of good land use.

To sketch this overall normative vision is merely to flag a long-simmering challenge, one that we have not over time faced very well nor, indeed, even thought about clearly. We may believe otherwise—that we in fact possess an overall standard of sustainability suitable for evaluating how well we are dwelling in place. But as critics have long asserted (to the small audiences who care to listen), sustainability suffers from various deficiencies, not the least being its vagueness. Too often, for instance, it merely refers to an ability to continue existing economic activities for the foreseeable future even as soils and waters deteriorate and biodiversity declines. Except in rare iterations sustainability hardly begins to draw together all of the normative factors that would seem pertinent to a first-class inquiry into good land use.

Here it is only possible to flag this third issue, dealing with our need to craft and embrace an overarching, normative vision of good land use. It is similarly necessary to leave under-developed the fourth of the foundation points, about the root, cultural causes of our environmental ills.

Our environmental problems are caused by our own behaviors. Indeed, it is useful to define an environmental problem, not as an unwanted condition in nature (a polluted river, an imperiled species), but instead in terms of the human behavior causing it (excessive farmland tillage, nitrogen spreading). To focus directly on the underlying human behavior is to invite pertinent question about causes. Why do we act toward nature as we do, for good and ill? Human behavior, it hardly needs saying, stems from many causes; it is no mean task to tease out causes of our

---

[8]The challenges of bringing health to the Upper Mississippi and other U.S. water systems are considered in Doyle and Drew (2012).

[9]I consider the issue at length in *Our Oldest Task*, particularly Chap. 8.

land-abuse. As a question and topic it receives too little attention, in part (one suspects) because it transcends academic disciplinary boundaries. Forgoing serious inquiry we too often latch on to simple models of human behavior. What environmental historians instruct—and they are likely our best scholars on the subject—is that interactions with nature are significantly influenced by dominant cultures, by the ways we perceive nature, value it, and understand our place it; similarly, by our time horizons (short term, too often), by our overconfidence in our cleverness, and by tendencies to discount or ignore nature's time-tested ways.[10] Culture does not account for everything—population and technology are vital also. But it plays essential roles, worsening environmental ills and resisting calls for mitigation even as it, in some ways, motivates good behavior.[11]

If these claims of historians (and others) are correct, then we are unlikely to constrain our misuses of nature without altering core elements of our culture, without, for instance, diminishing our senses of human moral exceptionalism, confessing our limited capacities, and appreciating nature's interconnections. We need to see ourselves as parts of complex communities of life. We need to recognize and confess that, much like other life forms, we depend for our flourishing on the healthy functioning and biological complexity of those communities. We need to categorize land and water uses, however undertaken, as matters of public business, not merely private concerns. It is our charge as value-creating beings to craft and then embrace sound normative standards to use as we struggle to live well in nature.[12]

The cultural reforms that seem needed for us to live well in place are ones that call into question key elements of modern liberal individualism.[13] (More on this below.) It is simply implausible to assume, for instance, that moral value resides in humans alone and that we enjoy this unique moral status as autonomous, self-interested individuals, as liberal thought often does.[14] In ways that are literally essential—that link to our creaturely essences—we are communal beings, embedded in human and natural orders. Necessarily, to live well is to behave as responsible community members.

---

[10]A leading and still-valuable work is Worster (1993), along with the same author's Bowl (1979). Other major studies include Steinberg (2009).

[11]I consider the cultural criticism of several major environmental writers in *A Good That Transcends*, Chaps. 1–4.

[12]My exploration of this point is in *Our Oldest Task*, principally Chaps. 3 and 4.

[13]Two particularly insightful backgrounds on liberalism are Fawcett (2014) and Siedentop (2014).

[14]The literature again is vast, including Singer (2011).

What It Will Take

Because we commonly do not expose to light or appreciate the cultural roots of our environmental ills we find it easy to seize upon simplistic solutions for them. Rarely do we ask whether the solutions we are considering would help remedy our cultural deficiencies—that is, address what we might usefully view as our true ills, underlying our environmental harms. Well-crafted environmental policies ought to do that, ought to confront root causes, just as they ought to work to move us toward our overall, normative vision of good land use.[15]

When we turn to our most thoughtful observers of our place in nature (Aldo Leopold, Wendell Berry, David Orr, and Pope Francis to name a few) we find them making, again and again, the same observations about the kind of culture we need to nourish.[16] For starters, they tell us, we need to appreciate nature as an ecologically functioning, intricate natural system or community, one that can operate in ways that are, for our long-term interests, better or worse. Nature is not simply a warehouse of physical commodities, a source of inputs for our economy. Perhaps a harder cultural challenge for us, as mentioned, is to infuse our natural communities as such with greater normative value and to sense an obligation to respect that value in daily living by sustaining the healthy functioning and biological complexity of the communities.[17] So prone are we to equate public welfare with the satisfaction of individual preferences that we have lost much of our ability to talk about the common good in other terms; to conceive it, and then pursue it, as a good that transcends the preferences of individuals acting alone.[18]

If we did give primacy in our normative thinking to the health and diversity of the land community, and if we did view that health and diversity as a prime element of the common good, then we could talk more openly about steps that might facilitate that good while clarifying the costs of needed tradeoffs. Many of those steps would require—they will require—that we think critically, and at various spatial scales, about our patterns of occupying nature. How might we modify our practices to sustain more ecologically sound water flows? What changes could help keep soil in place and sustain natural cycles of fertility? What many steps are required to increase the biological diversity of our landscapes, thereby fostering the many normative values furthered by the biodiversity?

To raise these questions is hardly to scratch the surface, of course. But the observations and questions take us far enough to lay out a number of further points relevant to climate change and human-drawn boundaries.

---

[15] I come to this conclusion in Chap. 8 of *Our Oldest Task*.

[16] All four are considered in *A Good That Transcends*, Chaps. 1–4.

[17] This was one of the central themes of Aldo Leopold's classic, *A Sand County Almanac and Sketches Here and There* (New York: Oxford University Press, 1949).

[18] A classic inquiry, still highly valuable, is Bellah et al. (1996).

- For people to live well on land (using land here broadly to include the entire community of life), they need to perceive healthy, vibrant land as a desired good and to evaluate, as questionable or wrong, activities that undercut the health and vibrancy of the land. This means, as noted, viewing land and resource uses as matters of public interest, not merely personal (landowner) choice.
- In all likelihood, this kind of concern about nature will depend upon a sense of social cohesion, some widespread sense that those who share a landscape are inevitably part of a single community.[19] Simply to live in a landscape is to forge ties with others who live there and to all other components, living and nonliving, of that resident natural community. With this sense of community membership needs to come the sense that good conduct includes the behaviors expected of a responsible community member, someone who tailors her decisions to help sustain the community's welfare.
- Good land use can hardly take place in the long run unless it is economically feasible for the individual land users.[20] They cannot do it if market competition forces them to employ degrading practices in terms of tillage, crop choices and rotations, chemical usage, harvesting plans, predator control and the like. This requirement can pose grave challenges when commodity prices are set largely by market forces that transcend and ignore local conditions. Market pressures, to be sure, can bring good consequences overall in the long-term. But they can also undercut sound land-use practices, even by well-meaning community members.
- As the above points imply, public discourse cannot simply be framed in terms of individual rights and individual preferences. This includes claims by landowners and others that they deserve compensation when they are called upon to act in ways consistent with land health. To argue this way, putting the individual above all, is to assume that good conduct is not a legitimate expectation of the community. It is to reject the idea that the community as such can properly set binding standards on its membership.
- Many of today's core cultural understandings are embedded in and fostered by our dominant social institutions, especially the market economy and private property. These institutions reflect and strengthen many of our cultural elements that encourage and legitimate bad land use; indeed, they make these harmful cultural elements seem both natural and inevitable.[21] Looking ahead, it would seem essential that we view such institutions clearly not as freestanding elements of the world but rather as human-crafted tools, ones that make sense only

---

[19]The theme has long been central to the writings of Wendell Berry.

[20]Again, the point is pressed often by Wendell Berry.

[21]The many cultural critiques of market capitalism, and its ecological effects, include Foster (2002) and Magdoff and Foster (2011). Nearly all writing on the issue builds on the classic work, Polanyi (1957) (original edition 1944). My critique is in Chap. 7 of *Our Oldest Task*. My comments on private property and the cultural of owning include *On Private Property: Finding Common Ground on the Ownership of Land* (Boston: Beacon Press, 2007).

when and so long as they foster the welfare of the community.[22] To the extent the institutions do not foster community welfare, then constraints should be put on them, on market activities and on actions by landowners.

At one point generations ago individual rights were understood as derivative of the community, property rights included: such rights existed when and to the extent their recognition fostered the common good.[23] Today it is often thought that rights play a different role, that they protect the individual from demands of the community, including demands to curtail actions that are plainly harmful to the common good. This line of rights thought is hardly without merit. But it has little place when it comes to our interactions with nature as such. There cannot be a legitimate property right to degrade the shared home of all; there cannot be a secure economic liberty that protects such behavior from challenge. Why would we create and honor such rights? The sticking point here really is a cultural and political one, not legal. Lawmakers have the flexibility to halt community-harming actions. In general, the rights of landowners best promote this element of the common good when they include an expectation that landowners will help sustain the healthy functioning of their home landscapes.[24]

- Local communities, of course, are embedded in larger orders. People living and acting at the local level need to recognize this, to see that they and their local lands help form larger natural orders extending across national boundaries and continental edges to the planet as a whole. With this recognition ought to come some felt sense of obligation to act as responsible members of these larger-scale communities. With it too should come the understanding that larger-scale communities must also remain healthy, with adequate places for people to live and flourish. This in turn can rightly translate into expectations and demands, established at that larger-scale level and reaching out to all peoples living within it. The few examples set forth above, of boundary-crossing problems, illustrate these points: wildlife migrations often cross whole continents (even more in the case of some birds); river-related problems can similarly extend a thousand miles and more. Climate change stands among these planet-wide ills, as do biodiversity declines and degraded oceans. Larger-scale communities, that is, can rightly press local peoples to act better.
- Finally, there is the matter of past mistakes and their consequences. Particularly over the past two centuries we have simply not lived on the planet in ways that can endure. In too many places and ways we have failed to succeed at what American conservationist Aldo Leopold termed the "oldest task in human

---

[22] A much-cited work on the benefits of embedding the market in our social and ecological orders is Daly and Cobb (1989).

[23] A useful introduction is Hunt (2007).

[24] I explore the issue at length in various works, including *The Land We Share: Private Property and the Common Good* (Washington, D.C.: Island Press, 2003).

history," to live on land without degrading it.[25] The consequences include much human suffering, as well as losses of soil fertility, species declines, dead zones, unnatural flooding, and more. Even as we look ahead, then, trying to craft better ways of living, we have the reality of existing degradation caused by past (and on-going) practices. They too are on the table for collective discussion and action. On one side, there is the desire and need to hold accountable those who have caused today's degradation. Injustice has taken place. But we have reasons to go slow here. The measures we take to do that, to redress past and on-going wrongs, need to be ones that are consistent with our much needed, land-respecting shifts in culture and politics. The top environmental priority, to reiterate, is to help everyone, everywhere, improve their uses of nature and embrace a culture that honors and sustains good land use. This means (among other things) avoiding whenever possible remedial actions that undercut this priority, even when taken to foster retributive justice.

A Future for Borders

These various statements and stances, conclusory but essential, supply a background for thinking about borders, about climate change, and about the plight of migrants. The starting point is to consider what it will take to help people live well in place, economically, politically, and culturally. Many of the needed components should be rather easy to list, however challenging they would be to bring about.

- It is hard to imagine people taking good care of their lands unless they enjoy a *sense of community* among themselves; unless they feel a kind of solidarity and civic-mindedness that motivates them to discuss the common good, to view uses of nature as public business, and to set expectations for behavior that respect nature.[26]
- Good land use, as previously noted, must be *economically feasible*. Market competition and related forces cannot be allowed to drive land-users by necessity to degrade. Further, this economic feasibility needs to be sufficiently secure—and users need to perceive it as secure—to encourage people to embrace a *long-term perspective* in their land uses. This will likely mean confidence that lands can be passed along to later generations, who can also survive if not flourish while using them well. The challenge here is substantial given global interdependence and may require, for instance, trade barriers to protect domestic producers.

---

[25]Aldo Leopold, "Engineering and Conservation," in Flader and Callicott (1991) (original written 1938).

[26]Individual landowners can and do, to be sure, tend their lands with care even in the absence of strong communal leadership on the issue. But many environmental challenges are ones that individual landowners acting alone simply cannot redress because they require coordinated action among many landowners. Examples here include reconnecting rivers to their floodplains and restoring and protecting large-scale wildlife habitat.

- This kind of community-centered commitment to the long-term will not help much unless land-users collectively in a region can *take charge of the landscape* to the extent needed to encourage good land uses and discourage bad ones. This means, in some fashion, power wielded at the collective level to contain unwanted land uses. Necessarily this includes the power to *define the meaning of property ownership* in terms of the rights and responsibilities that go with it.[27] This core need, in turn, implies a governance system that is responsive to community sentiment and not guided or overpowered by agents (domestic or foreign) interested solely in short-term profits.[28]
- The above requirements, especially those in the last bullet point, highlight the need for local people to be able to *resist outside forces* and interests that would push them in bad directions. The work of common-property scholarships has amply highlighted how users of a commons must be able to exclude outsiders who interfere with their governance systems; for instance, outsiders who manipulate local governance to facilitate their land degrading activities.[29] Even more, this would likely mean the ability to counter competitive market pressures that leave local land users economically unable themselves to act responsibly.[30] Free-trade practices have their benefits, but they come at great cost when they frustrate good land use at the local level. The biggest need for reform will likely be in international trade in commodities taken directly from the land, from farms, forests, and mines. Here is where free-trade rules can directly frustrate good land use while at the same time, practically speaking, limiting the ability of local governing institutions to design systems to reward good land use. To be sure, local power can be used in bad ways, harmful to the land, so accountability to higher levels of governance (as noted next) remains vital. But local people inclined to use their lands well need the freedom to do so. Too often today they lack that power.

These various points all have to do with steps to help local people engage in better land uses. But these local users, as noted, exist within larger scales, and it is important that they take into account the ecological needs of these larger scales. For these larger scales to be in shape—ecologically healthy, resilient, biologically diverse—local actors may well need to adjust their activities. We can again think of the migrating wildlife, of the rivers of Bangladesh, and climate change. Even as local people think of themselves as a distinct law-making community, out to promote local ecological health, they also need to see that they help compose larger-scale communities. This awareness, too, is a cultural element, already present here and there but in need of much growth.

---

[27] I consider the considerable flexibility of private property, and the options for reforming it, in both *The Land We Share* and *On Private Property*.

[28] Good inquiries include Eckersley (2004), Morrison (1995).

[29] A penetrating critique is presented in Wood (2003).

[30] Some of the costs of market competition are covered in Frank (2011).

At larger scales—regional, global—sufficient power is needed to pressure local areas and their residents to consider the larger scales and step up to help promote them. This includes some sort of power to set minimum responsibilities and then enforce them. In the case of global problems, the sovereign power might properly reside at the global level, even as exercise of the power draws in regional bodies or networks to tailor duties to regional conditions. (The power need not be vested in a global government as such; it could remain in the hands of international bodies so long as they function effectively.) In the case of more regional problems—the rivers of Bangladesh, again—the bulk of the work might rightly be done at the regional level, at least if regional governance systems can be set up that do promote the shared good of all states without, as usual, deferring to the wishes of the more powerful states.

These points about local and larger scales, about cultural change and the needed powers of communities, offer guidance on the kinds of solutions that would seem most sensible for addressing widespread challenges. Regional and global actors should be assisting local people to use their lands well, not just helping them implement good land use practices but, going further, fostering the kind of culture and community needed to sustain good land use. It should hardly need saying that this vision of local integrity and stability is undercut by much of today's development and humanitarian "aid," which too often weakens local control and solidarity while promoting economic activities that degrade or overtax natural systems. It is similarly undercut by free-market visions and land-grabbing, big-scale international investments.[31]

Put otherwise, when good land use is raised high as the ultimate goal, then it is no longer helpful to evaluate overall progress by using GDP figures for a state or region.[32] (This is particularly true when GDP is elevated by harvesting nonrenewable resources and by using lands in unsustainable ways; also when a considerable amount of local production is simply whisked away to other places.) The standards of evaluation should be quite different, framed in terms of good local land use and the nourishing of land- and community-respecting cultural elements.

This brings us around again to the issue of past degradation and accountability for it. It brings us also to the issue of climate migrants, who count among the victims of past and on-going degradation. Those parts of the world that have played smaller roles in global degradation need to have their perspectives heard and appreciated. Those who have done the most to cause problems ought to bear larger burdens in addressing them. Still, to say this and to demand justice is not to justify public policies inconsistent with the above, land-respecting principles. Global justice needs to supplement the work of promoting good land use everywhere, not distort or inhibit it.

---

[31]Critical surveys of large-scale land acquisition include Liberti (2013), Pearce (2012).

[32]The many deficiencies of GDP, and proposals for reform, are presented in Daly and Cobb, *For the Common Good.*

The Plight of Migrants

So what might these various observations and policy stances mean in terms of climate-change migrants? A number of points seem pertinent.

- To rephrase a point just made, efforts to deal with migrants should not involve helping them through resettlement schemes or other measures that undercut sound local living. They should not be deposited in places and numbers where their presence undercuts that overriding goal. One danger is that too many outsiders can frustrate or inhibit local senses of community and identity. Local people need to have or gain the sense that local lands belong to them, in an important, collective way. This might well require a sense of bondedness among the local people, a bondedness quite possibly facilitated and strengthened by senses of ethnic or religious cohesion.[33] Local people simply must sense that they inhabit and are responsible for their own homelands and that their descendants will similarly own, control, and enjoy the homelands.
- A corollary of the last point is that migrants who form anything like a major flow cannot simply decide for themselves where they want to go. There can be no right to settle in the place of one's choosing. Of course preferences can be taken into account. But in terms of settlement locations other factors, those discussed above, should carry great weight.[34]
- This in turn colors any claim that might be made about rights possessed by migrants who are forced to move due to climate change or related ecological ills. Rights rhetoric enjoys its greatest popularity and influence in the affluent West; it is less respected if not openly resisted elsewhere.[35] Globally, it is unclear whether the moral frames of individual rights are gaining or losing ground, particularly when one looks beyond rhetoric to underlying conduct.[36] In any

---

[33]Of course such cohesion can come at social costs, and steps should be taken to reduce them and to encourage humanitarian aid. But attacks on ethnic and religious bonds that promote senses of collective responsibility can themselves carry heavy costs, too often overlooked. To degrade such bonds is to promote the continued fragmentation of society generally (in the U.S. context, see, e.g., Rodgers (2011), Aronson (2017)), weakening senses of community, frustrating collective action, and opening even greater space for the dominance of market forces.

[34]To say that migrants alone should not decide leaves open, of course, how decisions might best be made. The factors identified here do not justify giving local people veto power about their acceptance of migrants. They do suggest that decision-making needs to give serious weight to the effects of migrants on the abilities of local people to foster and act upon a culture that values healthy lands. Migrants, one might presume, will typically focus on themselves and their short-term needs. By them the orientation is understandable and justifiable. International efforts to help migrants, however, need to embrace broader and longer-term perspectives. The short-term focus on the self and one's family is precisely the attitude that yields land degradation.

[35]Even in the West, rights-rhetoric is often attacked for its potential to corrode social relations and senses of community and its overuse to address problems better considered in other terms. For instance, Ford (2011), Glendon (1991).

[36]Hopgood (2013).

event, many in the West do find value in comprehending the moral status of migrants in terms of rights, however vaguely defined. However much weight one gives to the rights-based moral frame, it seems most sensible to understand rights as claims made against the international community as such. For varied reasons it is less desirable for migrants from one country to claim rights vis-a-vis another country or people within that country. Global problems are globally caused. Some states, to be sure, have played greater causal roles in bringing them about. But it should be the global community as such that passes judgment on the differing levels of state responsibility. It should be the global community —NGOs as well as intergovernmental efforts—that prescribes the extra effort required of a nation that has disproportionately brought on our environmental ills.

Liberalism and Borders

In the summer of 2015 Pope Francis released his encyclical, *Laudato Si'*. More than most reviewers noted, the document presented a wide-ranging critique of modern culture, not just a call to action on environmental issues and poverty. "Many things have to change course," he asserted, "but it is we human beings above all who need to change. ... A great cultural, spiritual, and educational challenge stands before us, and it will demand that we set out on the long path of renewal." (¶ 202) The cultural attack leveled by the Pope included within its scope a good deal of modern liberal thought, in its political leftist forms as well as (and even more) its expressions on the political right. Along with calls for greater humility, personal regeneration, and a broadening of moral values came a potent instruction to recognize interconnections and interdependencies, among people and between people and all other components of natural systems. The vision he sketched of our earthly plight, like that of Aldo Leopold before him, was one that embedded people within natural communities.[37] We are not chiefly freestanding, autonomous beings, Pope Francis tells us, even as we each possess moral value. We are not best understood either as liberated actors, free to pursue self-interest in the marketplace, or as rights-bearing actors routinely pressing demands on the state. To the contrary, we participate in larger orders, we belong to natural and social communities. And we are called upon, as community members, to act responsibly to uphold our communities, valuing our communities as such as well as other community members.[38]

This means, in the context of our lives in nature, that we will often need to subordinate our preferences and rights to the welfare of our communities, particularly the land community. All forms of liberal individualism would need revision

---

[37]I explore the Pope's writing, comparing with the work of Leopold and others, in Chap. 4 of *A Good That Transcends*.
[38]A good inquiry is Ophuls (1997).

were we to do as the Pope instructs: to respect our communities more than we do, to invest them with greater authority and align our activities with the common good. Given the late hour for key environmental ills, this kind of cultural shift cannot come too soon. In simplest terms, our need, in the context of our cultural ties with a nature, is for an appreciable shift toward more communitarian thought and away from individualism and human exceptionalism. The polestar should be this: Individual rights—property rights included—should contribute to the common good broadly defined and derive their legitimacy from that contribution.

Looking ahead we would benefit considerably if we demanded of ourselves and one another that our competing rights claims be reframed in just this way, if we insisted that anyone pressing forth a rights claim explain, as she does so, how the public's recognition of it would foster the broader common good. Those pushing forward claims of individual liberty, for instance, ought to admit up front that actions by one person almost always affect others, near or far, sooner or later. That being so, why should the liberty of one be respected when it harms and restricts the liberties of others? Why favor the negative liberty of the individual landowner who seeks to degrade nature over the positive, collective liberty of the community to keep its shared landscapes health?[39] The question, to be sure, may have a good answer, at least in the short term. But the question should always be raised and answers critically weighed. Rights claims pressed by the political left are similarly due for scrutiny in the same way. Indeed, a shift toward community, along the lines sketched here, could be every bit as disruptive to leftist thought as it is to neoliberal and free-market thought, especially when (as proposed here) the community is then pressed to promote particular values and land-use ideals.[40]

Viewed from the point of view of large-scale communities, the plights of climate-change migrants attest loud and clear that, collectively, we are failing in our efforts to live well on land. We have the science and technology to do much better. Fairly assessed and with a long-term perspective, economic considerations are also not a major obstacle. The sticking point is predominantly cultural and, therefore, political. Yes, people in wealthy countries need to show more compassion for the poor, particularly for those harmed by environmental declines. But the chief cultural need is not simply to elevate compassion for suffering individuals. Yes they are suffering. But their suffering is also an indicator of cultural ills, which means it is a symptom as well as a problem. To reach out to them with aid can conceal these symptoms while doing nothing about the deep-rooted cultural ills that have fostered their suffering.

Of course climate migrants deserve help. But that help should take place in a framework that attends to the overriding need to reform the modern world view. And with that cultural rejuvenation—that shift toward interconnection, community, broadened moral value, greater humility, and a longer planning horizon—should

---

[39]Useful background, particularly for American readers, appears in Kammen (1986).

[40]The tension in academic thought is explored in Mulhall and Swift (1992). A constructive synthesis is presented in Fowler (1999).

come major adjustments to such institutions as the market economy, private property, and territorial claims by political bodies.

This brings us back to the issue of human-drawn lines on the maps and how we might reconsider them in light of climate change and forced-migrations.

Easiest to see (though hardly easy to bring about) are needed changes in private property as an institution, both at law and, more importantly, in modern culture.[41] Property lines are among the most potent and culture-laden boundaries that we impose on nature. Property is best understood, not (as is common) as an expression of embodiment of individual liberty, but rather as a social institution, created and revised by the community to promote the collective good. Property is created by law and exists to the extent prescribed by law. In effect it delegates the sovereign police power of the state to individual owners, authorizing them to call in the police to restrict the liberty of other citizens, most visibly those whom property law labels trespassers. It is sovereign power at work here, the same power that might be used, for instance, to silence public speakers and to lock up undesirables. For starters, then, the reform of property needs to begin with a much clearer look at property as a social institution, at where property comes from, why it exists, how we might benefit from it overall, and why is has, for centuries, been challenged as a tool of domination, exploitation, and (increasingly) environmental decline.

Our understandings of private property need readjustment so that owners and users of nature curtail their degrading practices. That is unlikely to happen without a major shift in the ways we think about the institution overall; without a revived understanding of property as a human-crafted arrangement, rightly tailored to promote the welfare of the community.[42] The subject of property reform is too vast to develop here. In brief, however, the following reforms seem timely, all of which would make property boundaries more permeable:

- We need new *limits on using land and nature* to curtail degrading practices; a revival and overhaul, that is, of the long-entrenched legal principle that landowners should use what they own only in ways that cause no significant harm. At the same time, property law should empower owners to use their lands well and protect them when they do so. In the work of crafting sensible land-use norms, rural residents—all community members—should have full voice.
- Further, the much-needed land planning at larger spatial scales—planning, for instance, to reshape coastal zone activities, protect migration corridors, and reconnect rivers to their floodplains—will call for the relocation of countless current land users. This means an expansion of state powers of expropriation (or, as Americans term it, *eminent domain*). Here too we have in effect a weakening of property boundaries insofar as governments gain the greater

---

[41]As noted, my explorations of property as an institution include *The Land We Share* and *On Private Property*. I urge the conservation movement to take up the issue in "Taking Property Seriously," in *A Good That Transcends*.

[42]The writing of Joseph William Singer on the topic is particularly useful, including his (2000).

flexibility to take land from current owners, with compensation but without their consent.
- Particular attention should go to the protection of those parts of nature that we have traditionally owned in common (water, navigable rivers, wildlife) and to other parts of nature that are similarly vital to community health and that similarly transcend property lines—our atmosphere in particular.[43] Once again, the more such parts of nature are considered common property, the more we revise our understandings of property boundaries.
- Overlapping with the above three points but deserving of separate emphasis is the problem of outsiders arriving on the local scene and buying expanses of land and resources.[44] The practice might bring economic gains, particularly when calculated with no regard for the identities of winners and losers. But if local communities are to build and retain senses of community, if they are to take responsibility for their home landscapes, then they need to insist that such outside buyers become part of the local community and abide by well-crafted community standards and expectations.[45] It must be understood that to buy land in a distant place is to cast one's lot with the people of that place and be expected to assist in—even better, to affirmatively promote–efforts to keep lands, waters, and systems ecologically healthy, diverse, and beautiful.[46]

With respect to our other major type of boundaries, our political jurisdiction lines, the directions of needed change are more varied, weakening jurisdiction lines in some ways while at the same time strengthening them. The aims of the changes, as outlined above, all have to do with (i) helping and encouraging local people use their lands and resources in ecologically and morally sound ways while also (ii) empowering communities at larger scales (regional, global) to insist that local decision-makers help meet the ecological needs and moral concerns of the larger scale.

- As noted several times, the many environmental ills that transcend political boundaries call for planning and governance at the scale of the particular problem. This means power and decision-making at scales that transcend jurisdiction lines, among and within states To the extent that governing power shifts upward to larger scales, the powers of local decision-makers decline and their jurisdictional boundaries thus become more permeable. The challenge here, to be sure, is immense, particularly when concerns rise above the state level to problems best addressed regionally and globally. In some settings negotiations among state actors could yield adequate solutions, without the need

---

[43]A stirring call to do so is presented in Wood (2014).

[44]The power of global corporations over many states is examined in Screpanti (2014).

[45]The challenges that arise in settings where central states are unwilling or unable to define and enforce property rights are taken up, in an African setting, in Joireman (2011).

[46]This would include decisions by law-making communities to head in more socialist directions, as proposed in such works as Williams (2010).

to create some higher-level governance organization. But in many settings it is simply not conceivable that state negotiation will suffice, particularly when the states involved have widely differing powers and the problems affect them unequally. Again we can draw in the rivers of Bangladesh, greatly threatened in the long run by water projects proposed by China and India. On its own, in its weakness, Bangladesh cannot hope to protect the kinds of river flows needed to avert massive ecological degradation and catastrophic human suffering.

- Among the needs for action at larger spatial scales will be guidance that curtails the ability of local regions (states, or parts of states) to compete in the global economy by taking environmental short-cuts, by sacrificing local nature in the hope of short-term economic gain. Here, too, we encounter a massive challenge, given the needs of many poor regions to develop economically. Yet, it seems unwise in the long run to give anything like full freedom to local people to make this tradeoff, to degrade local nature for short-term gain, even apart from claims that they are competing unfairly. To allow this is, of course, to encourage it to happen; it is to foster competition among poor lands to see which one is most ready to degrade its natural systems and assets. To allow it is also to give a green light when global corporations, development agencies, and sovereign-investment firms show up, proposing deals that would bring about massive degradation.
- Cutting the other way, in terms of jurisdictional borders and their permeability, we have the reasons mentioned above why local lawmaking communities need adequate power to keep their lands and waters healthy. As noted, good land use must be economically feasible. This might well require economic protection of local land users from outside competition. It might require, that is, restrictions on free trade and other actions—overt subsidies, for instance—that cut against the free-trade ethos. The presumption should be this: Each community should wield whatever power it takes to make possible the good, long-term use of their local lands by local people. Many measures may be needed: restricting commodity imports; curtailing sales and uses of particular chemicals, plants and animals (whether or not genetically modified); banning technologies that degrade hydrologic systems; and more. Yes, such measures could curtail economic development as commonly measured (mismeasured, critics would say). But the path is the right one, likely the only one, when we raise high the goal of living well in place, the goal of sustaining natural systems and biodiversity at local and broader scales. When we understand a local area as a kind of commons, managed by local users, we can usefully draw for guidance upon the literature of common-property regimes.[47]
- As for migrants, driven by climate change and pretty much any similar force, the above points provide a frame for thinking about them and about boundaries that keep them from moving. They illustrate the global reach of many environmental ills just as they similarly illustrate the types of problems that require collective

---

[47]For instance, Elinor (1990).

action at scales that transcend political borders. Regional and global action is needed to address the problem, just as it is needed to address related, broad-scale problems that transcend boundary lines. Collective responses at that broader level, the regional or global, might rightly call upon states everywhere to do their fair shares in responding to the problems, with fair share perhaps defined in ways that does pay attention to the different roles that states have had in creating the problems.

Yet, a world in which this kind of arrangement exist, this allocation of power and understandings about borders, is unlikely to come about unless we attend better than we have to the root cultural causes of our problems, just as Pope Francis and others have observed. Culture change needs to be job one. Along with it, as job two, must come changes in institutions to reflect healthier, more durable senses of value with greater emphasis, as noted, on interconnections, natural systems, communities as such, other life forms, and the long-term.[48] Measures to help climate-change migrants, and indeed the ways we simply talk about their plight, ought to reflect these better cultural values and frames. How can we best help migrants given the long-term importance of helping people everywhere live in ways that can endure?

The better road ahead would seem to be one in which we revise rather considerably our dominant moral frames, along with the institutions built on them. Out of a reformed culture could come clearer, alluring visions of humanity dwelling on this planet. Such visions could supply polestars and ways to measure our progress. We will not make good progress so long as we see ourselves chiefly as a collection of rights-bearing individuals. We will stumble if, in lazy fashion, we define the common good as simply a summation of individual preferences. Global and regional problems will linger and worsen when political boundaries remain impervious to the needs of larger spatial scales. Good land use will similarly elude us at local scales so long as property boundaries shield irresponsible conduct. And yet, local people will struggle to engage in good land use when they lack the tools to protect themselves from outside forces that, too often, care little or nothing about the fertility and diversity of their natural homes.

The plight of climate change migrants calls us today to reconsider many of our boundaries and gives reason to make some of them more permeable. But such migrants join a suite of problems stemming from our failures to live sensibly in nature. Beneath them all are cultural elements ill-designed to grasp and address these problems. The worldview that got us into our current mess will not get us out of it. Guided by better cultural frames we could make better sense of climate change, migrants, and related environmental challenges and, seeing more clearly, craft better ways to deal with them.

---

[48] A good source for possibilities is Weston and Bollier (2013).

References

Aronson R (2017) We: reviving social hope. University of Chicago Press, Chicago
Bellah RN et al (1996) Habits of the hearth: individualism and commitment in American life, updated edn. University of California Press, Berkeley
Daly HE, Cobb JB Jr (1989) For the common good: redirecting the economy toward community, the environment, and a sustainable future. Beacon Press, Boston
Doyle M, Drew C (2012) Large-scale ecosystem restoration: five case studies from the United States. Island Press, Washington, D.C.
Eckersley R (2004) The green state: rethinking democracy and sovereignty. MIT Press, Cambridge, MA
Elinor O (1990) The evolution of institutions for collective action. Cambridge University Press, New York
Fawcett E (2014) Liberalism: the life of an idea. Princeton Univ. Press, Princeton, NJ
Flader S and JB Callicott (eds) (1991) The river of the mother of god and other essays by Aldo Leopold. University of Washington Press, Madison, p 254
Ford RT (2011) Rights gone wrong: how law corrupts the struggle for equality. Farrar, Straus and Giroux, New York
Foster JB (2002) Ecology against capitalism. Monthly Review Press, New York
Fountain H (2015) Researchers link Syrian conflict to a drought made worse by climate change. New York Times, A13, March 2, 2015
Fowler RB (1999) Enduring liberalism: American political thought since the 1960s. University of Kansas Press, Lawrence, KS
Frank RH (2011) The darwin economy: liberty, competition, and the common good. Princeton, Princeton Univ. Press
Glendon MA (1991) Rights talk: the impoverishment of political discourse. The Free Press, New York
Hopgood S (2013) The endtimes of human rights. Cornell Univ. Press, Ithaca, NY
Hunt L (2007) Inventing human rights: a history. W.W. Norton, New York
Joireman SF (2011) Where there is no government: enforcing property rights in common law Africa. Oxford University Press, New York
Kammen M (1986) Spheres of liberty: changing conceptions of liberty in American culture. University of Wisconsin Press, Madison
Liberti S (2013) Land grabbing: journeys in the new colonialism. Verso, London
Magdoff F, Foster JB (2011) What every environmentalist needs to know about capitalism. Monthly Review Press, New York
Morrison R (1995) Ecological democracy. South End Press, Boston
Mulhall S, Swift A (1992) Liberals and communitarians. Blackwell Publishers, Oxford, UK
Northcott MS (2007) A moral climate: the ethics of global warming. Darton, Longman and Todd, London
Noss RF, Cooperrider AY (1994) Saving nature's legacy: protecting and restoring biodiversity. Island Press, Washington, D.C.
Oberman K (2017) Immigration and equal ownership of the earth. Ratio Juris 30(2):144–157
Ophuls W (1997) Requiem for modern politics: the tragedy of the enlightenment and the challenge of the New Millennium. Westview Press, Boulder, CO
Orr DW (2016) Dangerous years: climate change, the long emergency, and the way forward. Yale Univ. Press, New Haven
Pearce F (2012) The land grabbers: the new fight over who owns the earth. Beacon Press, Boston
Polanyi K (1957) The great transformation: the political and economic origins of our times. Beacon Press, Boston
Rodgers DT (2011) The age of fracture. Harvard University Press, Cambridge, MA
Schindler DW, Vallentyne JR (2008) The algal bowl: overfertilization of the world's freshwaters and estuaries, rev. edn. University of Alberta Press, Edmonton

Screpanti E (2014) Global imperialism and the great crisis: the uncertain future of capitalism. Monthly Review Press, New York

Siedentop Larry (2014) Inventing the individual: the origins of western liberalism. Harvard Univ. Press, Cambridge, MA

Singer JW (2000) The edges of the field: lessons on the obligations of ownership. Beacon Press, Boston

Singer P (2011) The expanding circle: ethics, evolution, and moral progress. Princeton Univ. Press, Princeton, N.J.

Steinberg T (2009) Down to earth: nature's role in american history, 2nd edn. Oxford University Press, New York

Welzer H (2012) Climate wars why people will be killed in the 21st century. Polity Press, Malden MA

Weston BH, Bollier D (2013) Green governance: ecological suriival, human rights, and the law of the commons. Cambridge University Press, New York

Williams C (2010) Ecology and socialism: solutions to capitalist ecological crisis. Haymarket Books, Chicago

Wood EM (2003) Empire of capital. Verso, London

Wood MC (2014) Nature's trust: environmental law for a new ecological age. Cambridge University Press, New York

Worster D (1979) Dust Bowl: The southern plains in the 1930s. Oxford University Press, New York

Worster D (1993) The wealth of nature: environmental history and the ecological imagination. Oxford University Press, New York

Chapter 14
Neoliberal (Mis)Management of Earth-Time and the Ethics of Climate Justice

Michael S. Northcott

Abstract In this chapter I will argue that present day forms of economic accounting and management are changing the public and private representation of costs and benefits of consumption and production activities, including those which impact on climate change and energy use. I These foster a culture focused on near-term quantitative targets instead of attending to the intrinsic goods of production and service activities. The resultant short-termist mentality has notable impacts on the ecological sustainability of public and private investments. In this study of faith-based climate activism I show that individuals and communities who commit to ecologically sustainable activities do so primarily not from an accounting frame of near term risks and benefits. Instead they act because of their knowledge of the impacts of climate change on already existing persons, including farmers in developing countries, or climate risks for their own children and grandchildren. This relational frame for responding to environmental risks arguably has more cultural power in fostering sustainability than the narrowly quantitative cost benefit frame fostered by economic neoliberalism.

Liberal and Neoliberal Economic Management

The origins of economic liberalism may be traced to the English Civil Wars in the course of which around[96]. 190,000 people were killed, which was 4% of the English population at the time. Thomas Hobbes lived through the war and concluded that the human condition was essentially a war of all against all. In *Leviathan,* Hobbes argued that persons are first and foremost individuals who are 'free' and unencumbered by social roles or rules and who contract together for certain purposes—and particularly to defend themselves and their property from violence—with a sovereign authority—the State. Hobbes called the State 'Leviathan' which in the

M. S. Northcott (✉)
University of Edinburgh School of Divinity, New College,
Mound Place, EH1 2LX Edinburgh, UK
e-mail: M.Northcott@ed.ac.uk

bible of the seventeenth century was the word used for 'sea monsters' or whales (Hobbes 1651). In the woodcut image on the frontispiece, Leviathan is represented as a giant male bearded king who holds a sword in one hand and a shepherd's crook in the other. Beneath the king and to the left side of the title are images of warfare including a large canon, a castle, rifles and a cavalry battle scene. On the right there is a church, a bishop's mitre, forks depicting the words 'spiritual' and 'temporal' and a court of law which includes advocates, a judge and a jury. Hobbes was the first to describe the sovereignty of the early modern State as combining two forms of power —violent force and pastoral power. His argument, that men and women contract together to deprive themselves of some of their original individual freedoms by submitting to the State because otherwise they are at war, remains highly influential.

The other highly influential version of the liberal creed occurred in the opening paragraphs of Adam Smith's *Wealth of Nations*. Smith argued that people throughout history and in every society have always tended to engage in relationships with other individuals beyond their immediate kin in the mode of trade, or 'truck and barter'. Smith argued that by pursuing their own economic interests, comparative advantage and the laws of supply and demand would combine to turn myriad individually profit-oriented actions into a spontaneous order which advanced the wealth of all. There is poor historical evidence for Smith's claim that in past societies trade and competition were the *normal* way in which individuals obtained food, shelter and cultural goods, just as Hobbes' experience of violent civil war was exceptional (Graeber 2011). But Smith's proposal that societies that facilitate the division of labour, comparative advantage, and market exchanges between small and medium-sized business owners will experience increases in collective wealth unmatched by centrally planned economies, is widely accepted.

In economic history, liberalism took the characteristic form of *laissez faire*, according to which the duty of the State with regard to economic behaviours is to minimally interfere with free trade between firms and individuals: the principal economic purpose of the State on this account is to prevent theft and to prevent market monopolies. The rise of this doctrine was however accompanied by a land grab by the State and landowners between the seventeenth and nineteenth centuries which rendered most English and Scots individuals property-less, and hence dependent on the willingness of an employer to pay them wages for their *time* at work in factories or on farms belonging to others. In the latter half of the nineteenth century *laissez faire* was modified into a more social democratic kind of liberalism according to which individuals who lacked the freedom from waged labour conferred by land ownership or other kinds of wealth were owed certain duties by employers. These included, at a minimum, regulation of hours of work, and of other working conditions that endangered the health of workers (Steinfeld 2001).

In the twentieth century, two world wars, and the intervening Great Depression, prompted the rise of centrally organised public services and economic interventions by governments, including strong legal controls on banks and other private corporations. After the large scale State mobilisation of the Second World War, governments shifted the focus of the planning mode from war-time to peacetime civil, economic and social activities, including what were perceived as essential industries and services such as education, energy, housing, transportation, sanitation, health and social security. But against this a small group of economic liberals in France, Austria, the United States, and Britain, led by Ferdinand Hayek, mounted a new defence of the earlier form of liberalism. They argued the rise of State involvement and planning of economic activities and service provision would suppress market freedoms and economic growth, and they pressed for the reassertion of the *laissez faire* doctrine of the nineteenth century. Over the next four decades this group coalesced around the 'Mont Pelerin Society', and influenced the development of economic thought in central banks and universities, and especially in Chicago, Washington DC, and London (Mirowski and Plehwe 2015).

In the 1970s the neoliberal revival of laissez faire was first tried in the sovereign territory of Chile after a military coup against the democratically elected government of Salvador Allende which was first suggested by the CIA at the behest of President Nixon who had tasked the CIA from 1970 to organise opposition to the appointment of Allende as the democratically elected President of Chile (Snider 2008, 30–1). It was then adopted, often in the midst of civil war, in Central American States, including Guatemala, Honduras and Mexico, and in 1980 onwards by the administrations of Ronald Reagan and Margaret Thatcher in the USA and Britain. Under neoliberal policies most public services were taken over by private companies and user fees raised, including for health care and higher education (Harvey 2007). Much manufacturing activity was off-shored in order to reduce the power of trade unions in traditional industries and transferred to developing countries with weak labour and environmental laws. The four neoliberal decades have not realised their economic promise. While growth in national GDP continued in the USA and the UK, apart from a short hiatus in 1973–4 linked to the OPEC oil crisis, real wages growth did not keep pace and economic rewards were increasingly mal-distributed to the top one per cent of earners. Under neoliberal economic governance, economic inequality, private debt and insecure employment grew (Piketty 2014). Hence under the strident banner of a resurgent liberalism, genuine 'freedom' for the majority of the population declined.

Neoliberal economists, corporate lobbyists and 'think tanks' successfully persuaded social democrats as well as conservatives to reject the key early innovation of social democratic liberalism, which was that where individuals lacking property need to sell their time to employers in order to survive, this use of their time ought to be socially regulated so that employment is not coercive or punitive, and does not cause ill health. Hence the first laws passed in the nineteenth century to control working conditions concerned hours of work. In the same period Victorian cities installed water and gas pipes, sewers, electricity and telephone networks, and constructed reservoirs, public hospitals, railways and educational institutions, either by public subscription or with local taxes (Ashworth 1968). The legacy of these early modifications to liberalism was an enduring one since there are no western states where work hours are not regulated. And in American and many European cities, nineteenth century drains, roads, railways, bridges, and public buildings still form key parts of economic infrastructure in the twenty-first century.

Private industry benefitted from the rise of municipal infrastructure, as it did also from State provision of universal education, clean water, public health and so on since this facilitated and underwrote economic activities. But it responded to the social regulation of working time by introducing a new 'scientific' mode of time management in the workplace, known as Taylorism after the man who first introduced it (Taylor 1911). Time, and productivity per hour, in the workplace became central sites of contest between managers and workers, and firms and trade unions. Even the application of time and motion to factory work was insufficient to stem the reduced returns on capital arising from increased labour and environmental protection in the 1960s and 1970s. The resultant reduced returns on capital investment prompted a systematic attack by neoliberal economists, and the corporations and governments they advised, on unionised work and environmental regulation from the late 1970s in an attempt to 'liberate' capital from democratically developed regulatory constraints and collective planning.

Employment within neoliberal countries since the 1970s tended to move towards new kinds of 'flexible' contracts in which workers were less likely to achieve a living wage with only one job. In the UK five million workers are now officially designated as self-employed, with no employment rights other than a minimum wage per hour, even though most of them work for large service companies. They include many who were formerly employed in secure public sector jobs whose positions were down-graded when public sector organisations were privatised in pursuit of the purported neoliberal economic strategy to 'shrink the State'. Until the 1970s what were then known as the 'professions'—including academia, engineering, journalism, law, medicine and veterinary medicine, school teaching, and social work—were governed by vocational training and professional bodies which between them generated a considerable degree of autonomy from commodification and top-down managerial control. But a central element in the economic reform agenda which rose to prominence in the UK and USA after 1979 was monetisation of performance in all workplaces whether public or private since monetisation was said to expose all activities to market pricing mechanisms and therefore to promote greater efficiency and rationality while, at least in theory, reducing the role of government and the State. So for example, school teachers' performance since 1979 has increasingly been measured by the performance of children in regular testing, and salaries varied accordingly, while academics' performance is measured by grants obtained, research outputs produced, and numbers of students successfully taught. The effects of this cultural change have been profound and are leading to the 'deprofessionalisation' of the professions and their gradual transformation into commodified services run for profit by private companies in which professionals have little autonomy or agency in the crafting of their performance (Clark 2005). The demeaning and stultifying effects of this process for professionals are powerfully described in Richard Sennett's ethnographic study of the technologically managed workplace in his book *The Corrosion of Character* (Sennett 1998). This has produced in the professions, as well as in factories and shopping malls, what some call a 'neoliberal subject' who has been trained to internalise the new temporal regime so that time is internalised as the enemy rather than unreasonable management pressures to perform according to quantitative and temporal performance targets (Davies and Bansel 2005). The societal consequence is growing work-related mental ill health, which is already at epidemic proportions in some neoliberal domains.

Neoliberal management through quantitative performance measures is also at the heart of international and national climate change risk management. The 2015 Paris Agreement, which emerged from the twenty-first Conference of the Parties to the United Nations Framework Convention on Climate Change, does not discuss the principal geophysical cause of climate change, which is fossil fuel extraction, and subsequent use in furnaces, engines, power generating plants, and cement factories. The phrase 'fossil fuels' is not used *once* in the Paris Agreement (UNFCCC 2015). Instead the agreement presents a quantitative measure of global performance in managing climate risk which is stated intention to prevent the earth from average warming 1.5 or 2 °C over pre-industrial temperatures. It invites national governments to submit regular audits of their national greenhouse gas emissions and indicate how they plan to reduce future emissions so as to limit future warming. As with other kinds of neoliberal management the focus here is not on the activities which either exacerbate or mitigate climate change: instead it is on quantitative targets which the performance of such activities are supposed to be aimed at in the near and mid-term future.

Neoliberals claim to believe that the welfare of individuals is best advanced when societies devote themselves to quantitative measures of business performance and production, rather than focusing on welfare itself. The same logic is applied to the earth. The Earth is given quantitative targets which she is supposed to conform to once the disciplinary rigours of performance targets have been sufficiently internalised by businesses, citizens and households. Neoliberals do not acknowledge that there are Earth System limits to the environmental costs that businesses and societies generate in the pursuit of growth in economic and monetised exchanges, just as they refuse that there are psychological health costs to neoliberal management and the internalised flexible working environment. There is no international agreement to limit the extraction and burning of fossil fuels because neoliberal management does not operate through collective deliberation but through the top-down enforcement of numerical near and mid-term performance targets.

Ethical Temporalities and Climate Justice

Neoliberal blindness to Earth System limits on environmental and social costs reveals a distinctive philosophical approach to time reckoning in neoliberal economics. At the heart of liberal political economy is a temporal calculus first adumbrated by Smith according to which individuals and firms which act in ways that increase their private gain in competition with the short-term interests of others may be said to act beneficently because 'providence' working through 'natural' economic laws ensures that their actions produce a longer term increase in the sum of a nation's wealth, and hence welfare. Before Smith philosophers had argued that actions that had the character of being directed towards private gain could not be described as beneficent. Smith's approach, also commended by David Hume, is known by philosophers as the hedonic calculus and it was taken up by utilitarians, including Jeremy Bentham and John Stuart Mill, who argued that an action should be judged according to the sum of consequences it produced in the collective balance of human happiness and suffering rather than whether it was intrinsically right or wrong according to a traditional moral code (Bentham 1789; Mill 1863). In contemporary economic theory this approach is known as the 'Kaldor-Hicks criterion' according to which actions by individuals, firms, or governments which are intended to produce increases in wealth are judged beneficial regardless of whether these increases cause reductions in welfare to some individuals or parties (Kaldor 1939; Hicks 1940). So for example an inner city motorway that speeds movements of business and private vehicles may be judged socially beneficial because it increases collective monetised wealth even though it increases pollution in inner city streets and hastens earlier mortality of inner city residents and road users from particulate and nitrous oxide pollution.

Climate science is economically and politically controversial because it challenges the hedonic calculus of conventional economics, and the related tendency of neoliberal economists to promote shareholder value and monetary accruals in company accounts as measures of economic value over real world and longer term qualities such as human health, natural beauty, clean air and water (Northcott 2013). If climate science is real it requires collective planning beyond the price mechanism since the conventional assumption of marginal utility economists is that a good once it becomes extremely scarce will be replaced, at the margin when it becomes prohibitively expensive, by other equivalent goods prompted by the rising price of the scarce good. However, climate stability is so large a feature of the Earth System, and is intrinsic to the endurance of around two thirds of presently existing species, that it cannot perform to the conventions of marginal utility (Ackerman et al. 2009). Once climate stability has gone, and up to two thirds of species are extinguished, there are no marketable alternatives that will substitute for the qualities of life that these once sustained. As the Ehrlichs memorably argued, it is as though in extinguishing species, humanity is popping the rivets on her only spacecraft (Ehrlich and Ehrlich 1981), though NNW some venture capitalists, such as Elon Musk, believe space exploration may be one solution to the problem and are investing considerable sums in attempting to reach Mars.

Advocates of climate ethics and climate justice argue that short term economic gains from activities that pollute the atmosphere are accrued at the cost of medium and longer term climate impacts which will harm the welfare of future people (Gardiner 2006; Northcott 2007). They argue that harms to future people from raised temperatures, strengthened storms and rising ocean levels will be so great that they ought to be set against contemporary representations of economic gains from activities that emit greenhouse gases, and especially fossil fuel extraction and use, cement making, and deforestation. However modern market economies have a number of in-built feature which makes it possible to discount future costs against near term benefits. The first of these is that accruals to shareholder value from investment and production activities are represented in accounting practices on a quarterly and annual basis which are very short-term value measures. Secondly, conventional company accounts do not include social or environmental costs from productive activities that are not directly met by payments in company accounts. If a consumer of a company's product has health problems arising from its marketing and sale of tobacco or of high fructose corn syrup, the health costs do not appear in a company's balance sheet but are met by the consumer, and, in those nations with publicly funded healthcare, by healthcare providers. The same is true for environmental costs. Thus legacy costs from surface and deep mines for fossil fuels and minerals are typically discounted so heavily that mining companies post bonds that cover only a fraction of the actual costs of remediation. This problem is magnified with climate change since there are few traceable economic or political connections between producers and consumers of fossil fuels and those who experience health and other kinds of welfare loss from extreme weather events caused by climate change. The tendency to discount future costs against present benefits is long-standing but arguably exacerbated by the fact that modern money is the most influential medium of exchange value measurement, and it is in the main debt-based and hence interest bearing. Where the medium of exchange is interest bearing, then over time the relative monetary values of present benefits to medium and long term costs changes and hence the costs are annually discounted by the interest rate. Neoliberals tend to be monetarists and hence argue for a high discount rate of future, including climate, costs (Weisbach and Sunstein 2008; Ellingwood and Lee 2016).

Philosophers such as Stephen Gardiner and Edward Page argue that the reason for the lack of remedial action on climate change is because of its long-run temporal effects, and that present generations by failing significantly to reduce growing greenhouse gas emissions have 'broken the contract' between present and future generations (Gardiner 2006; Page 2006). The idea of a contract between generations was given currency in political theory in the eighteenth century in response to the French Revolution by Edmund Burke who argued that the 'theft' of aristocratic lands by the revolutionary government in France failed to honour the legacy of past generations, and that by breaking a contact with past generations revolutionaries also risked failing to honour the welfare of future generations (Burke 1760). A related perspective may be observed in the romantic origins of the environmental movement in the eighteenth and nineteenth centuries. Coleridge, Ruskin and Wordsworth argued that the destruction of wild species, and of beautiful landscapes, in the pursuit of industrial development dishonoured past (or *roman*) values and ways of life (Albritton and Jonsson 2016). Analogously the arts and crafts movement also looked back, to medieval crafts and the gothic, as sources of a traditional aesthetics and craftsmanship with which they imbued their distinctive contributions to architecture, fine art, interior decoration, manufacture and stained glass.

Burke however did not argue for the preservation of the political status quo, nor the romantics for nature conservation and respect for beauty, from claims about what present generations owe to the future but to the *past*. They argued that political and industrial revolutions were endangering moral and transcendent qualities intrinsic to the good life—including justice, beauty and the sublime—precisely by disrupting long-established political, moral and environmental qualities in the name of progress. Derek Parfitt analogously argues that it is a philosophical mistake to make moral and political judgments in the present on the basis of the envisaged claims on present people of unborn people and future generations (Parfitt 1984). This way of thinking unduly privileges future people, and future time, over present people and time: moral duties are owed to those who inhabit and share the *present* temporal space of moral actors and who may either be harmed by their negligence or benefitted by their virtue. It is precisely the tendency to discount present harms to particular people against utilitarian estimates of the benefits that will accrue despite particular harms which neoliberalism magnifies. Its advocates argue for public policies and practices which dissolve traditional economic and political institutions, laws and regulations in the belief that economic management, shorn of custom and political deliberation, will better advance human welfare by liberating the provision of goods and services from evolved customary constraints, and democratic political deliberation.

Christianity and Intergenerational Climate Risk Management

Despite their near ubiquity in the twenty-first century, capitalism and the modern nation State are relatively recent features of human history. The United Nations, the principal sponsor of international efforts to resolve climate change, is just seventy years old. Until the Reformation in Europe, the principal domain in which contracts, governance, law, and new technologies were ordered by a political power beyond households, villages and cities was the Holy Roman Empire. There were regions of the Empire which had elements of what today are known as capitalism and the nation state: Venice and Genoa were city states in the late medieval era whose merchants traded extensively with other city states and in such inter-regional trades capitalism in its modern global market form finds its origins (Epstein 2000); England was a land area which unusually had been united into one governable entity largely because of its island character and Anderson argues it was the first of the modern type of nation (Anderson) But before the European Reformation there was no formal international order of exchangeable capital, nor a means for addressing interstate problems other than war, or the authority of the Roman Catholic Church as exercised through the extensive land holdings of its monasteries and its influence over city authorities, and sovereign families.

Religions in their modern global form have their origins in the period Karl Jaspers identified as the Axial Age which is 2500–1500 years before the present. Of all the religions birthed in that era, Christianity had the most shaping power in the origins of modernity. Christian belief in the law-like and predictable governance of the physical cosmos underwrote the emergence of early modern science and the empirical method; Catholicism as an international worshipping community underwrote trust between distant traders in cities where early forms of international trade and capitalism emerged; and Christian monks invented some of the key technologies which shaped the modern world, and the modern relation to nature, including the clock, the windmill, the deep plough, and monks and clerics sponsored the large scale copying and reproduction of texts and especially the bible which fostered literacy.

Lynn White Jr. in a famous paper argued that this shaping power of medieval Latin Christianity over the natural order was a key historical root of the modern ecological crisis and in so doing White was the first to argue that religion plays a key role in shaping the human use of the environment (White 1967). The Reformation however, while it underwrote some features of the medieval inheritance—and especially science and technology—significantly changed Christian culture in Protestant regions. Until the Reformation the driving spiritual concern of Catholic religion for at least a thousand years had been the mediation of salvation in the *next* life. A principal ecclesiastical cause of the Reformation—leaving aside arguments about external causes such as the rise of mercantilism and the early modern nation state—was the extent to which the penitentiary system of good works and penance had come so to dominate Catholic Christian culture that it seemed to Reformers such as Martin Luther to have subverted the emphasis of the founders, as recorded in the Christian scriptures, on salvation through the forgiveness of sins as mediated by Jesus Christ. The system of indulgences, good works, and memorials to the saints which had become so prominent by the sixteenth century in the Latin economy of salvation seemed to the Reformers to represent a form of bondage to priestly power which was a contradiction of the original conception of salvation as freedom from debts as indicated in so fundamental a Christian text as the Lord's Prayer. By sweeping away this culture of penance, and the strong emphasis on salvation in the next life, the Reformers gave to Christianity a new this-worldly focus which ultimately led in Protestant culture to a new emphasis on the meaning and quality of everyday life, and to progress in material, political and scientific advancement in that life, rather than on the progress of the soul from this life through the penitential system to the life hereafter (Taylor 1989).

The transformation of Christian culture in the Protestant Reformation also significantly changed social perceptions of risk and danger. Everyday concerns such as having enough food to put on the table remained prominent in the daily life of most people. But the Reformation significantly shifted the balance between such concerns and concerns about the salvation of the soul in the life hereafter. Hence after the Reformation far less was invested collectively in religious monuments and sacrifices, such as the building of great cathedrals, arduous pilgrimages to holy places, and the expensive decoration of shrines to the saints. Instead individual and social surpluses from economic activity were increasingly invested in more this-worldly and secular institutions such as city and guild halls, agricultural and housing improvement, and the arts and sciences.

Risk management in Catholic culture had a significant intergenerational element because the ways in which the risk of damnation was managed was closely linked with the intergenerational cult of the saints. From Augustine of Hippo to Thomas Aquinas and up to and including Pope Francis, Catholics believe that by doing good and virtuous works, by honouring the saints of the past, and by calling on the prayers of the saints in heaven, these will speed their own souls from earth to heaven at the end of their mortal lives. Protestants mostly no longer believe this or think like this. If there are duties owed to the living in terms of moral actions these are best fulfilled by following the moral law rather than by being over-concerned about the future state of one's soul. If there are duties owed to the dead, these are primarily paid by an honourable funeral and burial. Priests do not need to be paid to say masses for the dead for months or years after they die. Neither do the bones of the dead need to be preserved in or near churches, awaiting the resurrection of the last day. Cremation of the bodies of the dead even becomes standard practice among many Protestants by the late nineteenth century.

Arguably then, intergenerational responsibility becomes a harder concept to underwrite in Protestant than in Catholic cultures. But from the first glimmerings of the environmental movement, which began, as did the industrial revolution, in Protestant cultural contexts, intergenerational responsibility was appealed to. The first recorded modern environmental protest occurred in the English Lake District when the City of Manchester proposed to raise the level of Thirlemere Water to turn it into a water storage reservoir to meet the needs of the city (Ritvo 2009). Ruskin and others who opposed the reservoir argued precisely that it would be a betrayal of the present generation's responsibility to future people who would no longer be able to enjoy the amenity of the valley in its pre-industrial state. Ruskin made a similar argument about atmospheric pollution from the chimneys of Manchester and other industrial towns when he said that it would prevent artists in the future from painting beautiful sunsets and vistas of the kind that J.E.W. Turner had painted (Ruskin 1884).

As we have seen, a number of modern philosophers, like Ruskin, argue that the long-run and irreversible climatic effects of atmospheric pollution are a cause of intergenerational injustice and that failure to prevent climate change represents a break in the covenant between present and future generations (Gardiner 2011). Given the role of religion in sustaining intergenerational awareness in the past, a number of Christian theologians have argued that Christian culture, and particularly Christian congregational life and worship, has a potentially valuable role to play in promoting greater intergenerational awareness of the risks of climate change and the need to mitigate them, including Northcott (2007), Muers (2008) and Jenkins (2013). To test out the idea that Christian culture might sustain intergenerational ecological awareness Northcott led a research project at the University of Edinburgh in 2013–16 entitled 'Caring for the Future Through Ancestral Time. The research aim was to discover whether, and which, perceptions of connections between the present, the past and the future play a role in forming contemporary environmental awareness and behaviours among Christians, and particularly among Christians involved in efforts to promote ecological responsibility among a cohort of church-goers whose congregations belong to a network called Scottish Ecocongregations. The purpose of the network, which received some funding from the Scottish government, is to raise environmental awareness among churchgoers, and to assist church communities and members in practising environmental behaviours such as recycling, reducing energy use by promoting 'low carbon behaviours', and gardening for biodiversity and to grow local food. Four field researchers visited 20% of churches in this 340 strong network of churches, conducting interviews, attending services and other meetings, and investigating the environmental activities and messages that the network promoted. Researchers used mixed methods broadly of an ethnographic nature including participant observation; unstructured and semi-structured interviews; group discussions; documentary and internet discourse analysis. Researchers also gathered comparative data from other environmental activist groups and communities, including Transition Towns in Scotland, and faith-based environmental activists in England, continental Europe, and North America.

Researchers found many kinds of temporal signals and symbols in the buildings, environments and ritual spaces of Scottish Ecocongregations. These include long-lived lichen on stone graveyard monuments; lists of serving parish clergy in some cases going back to the pre-reformation era; many church towers have large analogue electrically driven clocks; some churches ring bells to mark the commencement of the Sunday worship meeting; most churches display lists of names of parishioners killed in the 1914–18 and 1939–45 World Wars; some churches display lists of children on the church baptismal roll, and dates of baptism; most church buildings have gravestones marking burials of past members and parishioners in graveyards around the church; many church buildings have memorial stones or plaques honouring the names and dates of dead persons, often patrons or local landowners or clergy, on the floors or walls of church interiors; a few Ecocongregations—such as the Iona Community—worship at sites which have memorial stones and other archaeological remains going back to the first millennium of the Christian Era in Scotland; some churches are named after famous individuals—or saints—in Christian history or have plaques, stained glass windows, tapestries or other symbolic depictions in which the lives of the saints are depicted; all churches use the sign of the cross at various points in their interiors, a sign which represents the means of death of the founder of Christianity, Jesus Christ, at the hands of the Roman Empire almost two millennia ago, in 33 AD.

To these more explicitly religious temporal symbols are others which are more 'secular'. A number of churches, particularly those situated in rural areas, have an oil tank either in the churchyard or in a purpose-built room somewhere inside. The oil tank is a temporal marker of a different kind. It contains a reserve of black liquid which is burned in a furnace to boil water to heat the church through hot water pipes and radiators under church pews or on the church walls. This fossil fuel store represents the millennia old storage of sunlight by plants and shellfish from the Pleistocene era and earlier. This store of ancient carbon is slowly released into the atmosphere during the winter period to keep the church warm during worship services. A few churches in the Ecocongregations network have chosen to replace, or at least supplement, fossil fuel heating with renewable energy infrastructure. Selkirk Parish Church, which is a historic church in the Scottish Borders, has installed a large array of solar panels on its roof. In the church interior is a display which gives two temporal readings: one shows the amount of energy being produced in the present moment, the other how much CO_2 the solar panels have displaced by producing renewable electricity in their lifetime. Another church in the Scottish Borders with an oil tank by the church entrance also displayed on its notice-board a Venn diagram and statistical tabulation of the annual carbon footprint of an average church-member based on surveys conducted by a member of the Ecocongregation committee.

Churches are ritually involved in the marking of time through their patterns of worship. Most worship services, and especially in Protestant churches which formed the majority of subjects in the project, are held once a week on a Sunday, the seventh day of the week, which is also traditionally in Western societies a work and public holiday. Sunday is said to be the day on which Christ rose from the dead and hence the weekly gathering of Christians for worship was moved from the Jewish Sabbath to the Roman-designated 'Sun-day' in the first Christian century. The seventh day in the shared Jewish and Christian creation story in Genesis also marks the day on which God is said to have rested from the divine work of the creation of the earth and of life. Hence it is traditionally both the day of rest from work, and the day set aside for divine worship in which the creator is honoured and remembered. Most churches in the study also organise their worship around seasonal markers from the pre-Reformation liturgical calendar which mapped crucial events in the life of Christ—and in particular Christ's birth, death and resurrection—onto the climatic and hence agricultural seasons of the Northern Hemisphere. Christmas Day is celebrated at the time of minimal insolation, the Winter Solstice, and Easter Day is celebrated at the formal commencement of Spring, which is traditionally said to begin with the full or 'paschal' moon after the Spring Equinox, when the hours of daylight are longer than the hours of darkness and the increase in sunlight coaxes leaves back onto trees and seeds to sprout in the soil. This mapping of worship onto the earth year in the Northern Hemisphere is also repeated in some other regular ritual events such as Harvest celebrations when gifts are gathered, more often now from supermarkets than gardens and fields, for charitable distribution and the fruitfulness of the earth is celebrated as divine blessing.

Given many strong temporal markers in Christian ritual and church buildings, together with regular rehearsal of the events of the life of Christ two thousand years ago in scripture readings and hymn singing, project researchers envisaged that members of Ecocongregations might display heightened awareness of legacy in their actions with respect to the environment and energy use. We also envisaged that given the long-term and intergenerational character of church organisations they might be better at balancing short-term and long-term costs and benefits in planning for environmentally sustainable infrastructure investments compared to private companies. To investigate the first claim we conducted forty in-depth interviews in which the environmental behaviours, interests and motives of ecocongregation members were discussed.

We found two strongly temporal inflections on environmental awareness among my interviewees. The first we call 'descendant time', and the second 'presentism'.

A majority of interviewees in our sample were over sixty years-old. This reflects the fact that churchgoers in Scotland are older than the general population. They also therefore often have grandchildren as well as children. A number of interviewees referred to the horizon of environmental risk in the daily news and its potential impact on children and grandchildren as a major motive for individual and church engagement in efforts to reduce energy. For example, one interviewee states:

> I think it (ecocongregations) is an aspect of the church and it is very topical at the moment. Every time you pick up the paper it was about global emissions and if you can believe it 'we are all doomed'. You can be flippant about it but it is a serious issue and we owe it to our children and grandchildren to be sensible and not be burning lights all the time (Scottish Borders Ecocongregation member).

A second interviewee talked more positively about planetary care as a response to the horizon of climate risks:

> We have to look after this planet. We are trying to look after a planet that our grandchildren are going to live in and to ensure they are not going to get flooded, or have drought: there are wars over water, the sea level is rising round about Scotland, people can see there is climate change (Glasgow Ecocongregation, Minister)

And a third, this time in a group discussion, made the connection between past and future through the retelling of the Christian story:

> Every year we compile a service that is used at the other churches in the group...and that was one of the things we pushed that was the title of one service, that small things can make a difference. There is a passage in our most recent service again emphasising this message and talking of stories in the past emphasising the effects that our actions now will have on our children and grandchildren. That is one of the things that captured our imagination. (Midlothian Ecocongregation group discussion)

The other notable element in this group discussion was the recognition that an Ecocongregation could encourage its members, and the community in which it is set, to engage in pro-environment behaviours which individually seem small but which, when replicated in a community, acquire greater cultural force, as well as adding up to a larger collective impact of humanity on the environment going forward. That myriad small actions quantitatively add up into a larger collective impact on the planet, either negative or positive, is as much a spatial as a temporal concern. Six plus billion people inhabiting the limited space of a finite earth impact much more on planetary space, including the capacity of atmospheric and oceanic space to absorb greenhouse gas emissions, than the one billion people who were alive at the beginning of the twentieth century.

The spatial theme was evident in other interviews which stressed the importance of being aware of environmental impacts on people who spatially live far from their source but who, through the unitary space of the Earth System, are in effect planetary neighbours:

> It is about loving our neighbour. I preached at the creation care service, and said then we need to think about people we don't know as neighbours. People far away. But also our great grandchildren who are not born yet (Retired clergyperson, Ayrshire Ecocongregation).

The distant and hence hidden spatial effects of environmental impacts has been a concern of faith groups for some time, and in relation to toxic waste it led in the 1980s to the generation of the faith community-originated concept of 'environmental justice'. This concept emerged out of a study funded by the United States United Church of Christ's Commission for Racial Justice into the location and population effects of toxic waste dumps in the United States. The study found that these dumps were almost exclusively located in neighbourhoods close to the residencies of people of colour. Hence pollution of ground water and air from these facilities, as well as noise and pollution from truck movements associated with them, therefore disproportionately affect low-income people of colour (Chavis and Lee 1987).

Related analysis emerged in theological literature in the 1990s on postcolonial relationships between developed and developing countries, in which it was observed that developed countries after the formal end of colonialism used developing countries as sites for the extraction of raw materials for their industries including fossil fuels, rare metals, rare earth and uranium, and as sites for dumping of hazardous wastes and post-consumer waste. Extraction sites in developing countries are typically managed in much more hazardous ways, which inflict health and environmental problems on local employees and communities, of a kind which are outlawed by employment and environmental legislation in developed nations. For example, the present writer observed the contrast between the environmental pollution of the operations of the Shell oil company in the Niger Delta compared to North Sea offshore oil and gas fields (Northcott 1996). In 2007 I extended this postcolonial analysis of international environmental justice and injustice to a discussion of the effects in developing countries of climate change in terms of 'climate justice', a phrase first coined by the climate change unit of the World Council of Churches in (Northcott 2007).

The concept of climate justice was taken up by the UK church-based charity Christian Aid whose Scottish director Kathy Galloway addressed the annual gathering of Scottish Ecocongregations in 2015. She began her talk with a quote from an African-American writer, Harris Walker, which she described as her personal mantra:

> Love is not concerned with whom you pray or where you slept the night you ran away from home. Love is concerned that the beating of your heart should kill no one.

For Galloway climate change is an urgent development and faith priority because atmospheric emissions from fossil fuels are already impacting on the lives of very poor people living close to the edge of survival as nomadic herders in Northern Kenya, or farmers in Malawi. Christian Aid made climate justice a central aspect of their communication strategy with supporters because their partners in the developing world tell them that climate change is already impacting on the lives and livelihoods of those Christian Aid and their developing country partner NGOs are trying to assist. A retired member of an ecocongregation made the same connection, but in a more personal way:

> When we go to Malawi we help people there design and install solar (thermal water heaters) using tin, black paint, glass and local waste materials. We give regularly and are in daily contact with people in Malawi installing solar (photovoltaic) panels. We put solar on our own house, both thermal hot water and two kilowatt PV panels, and we had drilled a 100 m deep bore hole for ground source heat pump which cost £9000, plus the drilling. I dug the square meter hole myself and the trench to the house for the pipes but once installed it is maintenance free and we get the electricity from 'Good Energy' so all renewable. It is a blessing to live in a zero carbon house. We don't even need the heat pump in the summer as we have enough hot water and electricity from the (PV) panels (Ayrshire Ecocongregation member).

Malawi came up five times in the fifteen interviews I conducted. This not only evidences a long-standing relationship between Scotland and Malawi begun in the nineteenth century heyday of the British Empire when many people from Scotland served either in the armed forces, or in colonial services including agriculture, education and healthcare. It also evidences that church communities are motivated towards action on climate change less from reflecting on long-term environmental impacts on the planet and future generations than from an awareness through global Christian networks of the *present* challenges to peoples' livelihoods represented by climate change.

Three interviewees suggested that this *presentist* sense of the damage climate change is already doing is the driver for many of those in faith communities involved in efforts to promote more just and sustainable relationships between developed and developing countries, including not only on the use of atmospheric space for fossil fuel wastes, but other economic and trading relationships:

> I have a hard time perspecting (sic) anything into a long term scale. When it comes down to it I think about what can be changed in the here and now which has a short term effect to make things better. That influences me more than thinking about the long term, if I ever stop to think about that. Like choosing to purchase fair trade coffee over non-marked coffee because 'now' there is one more tick in the box and the store might be influenced to invest more in the fair trade side of things. I have done my research on the fair trade coffee thing so I know what that choice means further along the line beyond me. It is not necessarily a time influence. It is more who is employed on the other end, whom am I supporting now. I won't say I am not aware of transport, food miles etc. and how this influences the planet as a whole. I am aware of it. I am not happy about it. But I think there is always time to do the right thing. I don't know if we will run out of time. The kingdom begins today so that is more positive. (Faith-based conservation NGO worker, Edinburgh)

A group of young people involved in campaigning for the church to divest from fossil fuel shares had the same sense of presentist urgency:

> For me right now it is that we should stop burning carbon sooner rather than later, specially when we know we have got 5 times the amount of oil that we can safely burn, we did a lot of research on it, numbers etc. It is 285 months now the earth has been above average temperatures which is longer than we have been alive. And that put it into more perspective for us the fact we found that out. Even what most people think is norm is abnormal (Group discussion with Fife Ecocongregation youth group)

A minister in a Scottish Borders church made a very interesting observation which underwrites and helps to explain this ethical sense of the immediacy of the problem and the importance of immediate action towards its resolution. He argues that presentism is connected with an inability to relate to the apocalyptic scenarios of a significantly climate-changed planet such as those often communicated in climate science and secular environmentalist campaign literature:

> We don't really have a concept of the future - we can't conceive of the world being a sea-change different so when you try and explain that sort of thing to people they find it confusing and upsetting and not very enlightening. People are much better at dealing with problems that are here and now so I find that sort of tactic - you know the sky's falling in - does not work. But what they are good at is that sense of righteousness, of ethical behaviour now, that it is a good thing to do doing things because it is good, not because it has good results, and people are sensitive to that (Borders Ecocongregation Minister).

Conclusion

I began this paper with a consideration of the ethics of temporal relationships in neoliberal economic management. I showed how neoliberalism underwrites and deepens the tendency of mainstream economic liberalism to promote increases in the economic utility of individuals and firms as intrinsically good over actions directed towards more traditional conceptions of the human good such as compassionate action to reduce the suffering of others, and actions which increase the beauty as well as the productiveness of the Earth. Such actions, though traditionally considered intrinsically good by classical Greek philosophers and religious teachers including Christ and Buddha, are increasingly set aside by the short-term hedonic calculus of modern economic liberalism, and especially in its more extreme neoliberal form.

The data from our research into faith-based environmental activists indicates that the neoliberal tendency to favour economic utility, and performance targets which measure utility in business, citizen and public practices, over intrinsic goods is resisted by such activists because they prioritise compassionate action to reduce the suffering of others, and especially that associated with environmental injustice. The grounds for their resistance are various but among the strongest motives for this resistance are three. First faith-based activists argue that they have particular duties to hand on the planet in at least as good a condition as they received it to their children and grandchildren. This I call descendant time. Second, they argue that they have particular duties to do something to ameliorate the suffering caused by climate change, and other environmental impacts of economic activity and international trade, on poor developing world farmers, fishers and others, and that they have a duty to do this now. This I call presentism. Third, and related to the second point, they argue that there are spatial as well as temporal implications to the global as well as local environmental impacts of economic activity and infrastructure which are not sufficiently captured in conventional familial or national conceptions of moral and political duties. For faith-based environmental activists, climate change is already impacting the lives and livelihoods of people living in the developing world who have had least to do with its causation. Wealthy nations, corporations and householders therefore ought to 'do the right thing' by ending their reliance on fossil fuels, and ending forms of trade and economic activity which refuse that the developing world can be used as a dumping ground, or an extraction or employment zone, where the footprint of fossil fuel use or other kinds of economic activity can be hidden from view.

Faith-based prioritisation of the intrinsic good of actions directed towards reducing the present and known suffering of others, and reducing human impacts on the Earth System, provides an important corrective to the dominant neoliberal framing of climate change risk management around mid-term and future quantitative atmospheric performance targets. Neoliberal economic management legitimates actions known to be harmful to other people and the environment by situating them in a political and economic calculative nexus which gives the appearance that risks incurred in the present by doing the wrong thing will be manageable in the future with the aid of performance targets and surveillance of the performers. Neoliberal subjects are familiar with these strategies in the work place, and even when they turn on their smart phones. The Earth as neoliberal subject may prove less tractable, no matter how many temporal performance targets and surveillance systems she is subjected to.

My research finding lends empirical weight to Parfitt's argument that moral choices and political judgements ought not to pit the interests of unknown future people against the interests of really existing persons, and it goes against the claim of Gardiner and Page that the primary reason for failure to act on climate change is an inability to give sufficient moral and political weight to future people. On the contrary it is precisely because faith traditions and narratives—for example the Christian narrative of the Good Samaritan—argue for the moral priority of reducing suffering in the present that they underwrite individual and collective actions to mitigate climate change *now*.

References

Ackerman F, DeCanio SJ, Howarth RB, Sheeran K (2009) Limitations of integrated assessment models of climate change. Clim Change 95:297–315

Albritton V, Jonsson FA (2016) Green victorians: the simple life in Ruskin's Lake district. University of Chicago Press, Chicago

Anderson B (1983) Imagined Communities: Reflections on the Origins and Spread of Nationalism. London, Verso

Ashworth W (1968) The genesis of modern british town planning. Routledge and Kegan Paul, London

Bentham J (1789) Introduction to the principles of morals and legislation. Oxford University Press, Oxford

Burke Edmund (1760) Reflections on the revolution in France. J. Dodsley, London

Chavis BF Jr., Lee C (1987) Toxic waste and race in the United States: a national report on the racial and socio-economic characteristics of communities with hazardous waste management sites. United Church of Christ Commission for Racial Justice, Washington DC

Clark C (2005) The deprofessionalisation thesis, accountability and professional character. Soc Work Soc 3:182–190

Davies Bronwyn, Bansel Peter (2005) The time of their lives? Academic workers in neoliberal time(s). Health Sociol Rev 14:47–58

Ehrlich Paul R, Ehrlich Anne H (1981) Extinction. Ballantine, New York

Epstein SR (2000) Freedom and Growth: the rise of status and markets in Europe 1300-1750. London, Routledge

Ellingwood BR, Lee JY (2016) Managing risks to civil infrastructure due to natural hazards: Communicating long-term risks due to climate change. In: Gardoni P, Murphy C, Rowell A (eds) Risk analysis of natural hazards. Risk, Governance and society, vol 19, Springer, Cham

Gardiner Stephen M (2006) A perfect moral storm: climate change, intergenerational ethics, and the problem of moral corruption. Environ Values 15:397–413

Gardiner SM (2011) A perfect moral storm: the ethical tragedy of climate change. Oxford University Press, Oxford

Graeber D (2011) Debt: the first 5,000 years. Melville House, Brooklyn, NY

Harvey David (2007) A brief history of neoliberalism. Oxford University Press, Oxford

Hicks JR (1940) The valuation of social income. Economica 7:105–124

Hobbes T (1651, 1996) Leviathan. In: R. Tuck (Ed), Cambridge University Press, Cambridge

Jenkins W (2013) The future of ethics: sustainability, social justice, and religious creativity. Georgetown University Press, Washington DC

Kaldor N (1939) Welfare propositions and interpersonal comparisons of utility. Econ J 49:549–552

Mill JS (1863) Ultilitarianism. Parker, Son and Bourn, London

Mirowski P, Piehwe D (eds) (2015) The road from Mont Pelerin: the making of the neoliberal thought collective Cambridge. Harvard University Press, MA

Muers R (2008) Living for the future: theological ethics for coming generations. T. and T. Clark, New York

Northcott M (1996) The environment and christian ethics. Cambridge University Press, Cambridge

Northcott M (2007) A moral climate; the ethics of global warming. Orbis Books, Maryknoll NY

Northcott M (2013) A political theology of climate change. B. Eerdmans, Grand Rapids, MI, Wm

Page E (2006) Climate change, justice and future generations. Edward Elgar, Cheltenham

Parfitt D (1984) Reasons and persons. Oxford University Press, Oxford

Piketty T (2014) Capital in the twenty-first century. Harvard University Press, Cambridge MA

Ritvo H (2009) The dawn of green: Manchester, Thirlemere, and modern environmentalism. University of Chicago Press, Chicago

Ruskin J (1884) The storm cloud of the nineteenth century: two lectures delivered at the London institution february 4th and 11th, 1884: Digital edition Project Gutenberg, 2006. http://www.gutenberg.org/ebooks/20204

Sennett R (1998) The corrosion of character: the personal consequences of work in the new capitalism. W. W. Norton, New York

Snider LB (2008) The agency and the Hill: CIA's relationship with congress, 1946–2004. Center for the Study of Intelligence, Central Intelligence Agency, Washington DC

Steinfeld RJ (2001) Coercion, contract, and free labour in the nineteenth century. Cambridge University Press, Cambridge

Taylor FW (1911) The principles of scientific management. Harper and Brothers, New York

Taylor C (1989) Sources of the self: the making of the modern identity. Cambridge University Press, Cambridge

UNFCCC (2015) Adoption of the Paris agreement: proposal by the president: draft decision -/CP21, 11 December 2015 at https://unfccc.int/resource/docs/2015/cop21/eng/l09.pdf. Accessed 23 Aug 2016

Weisbach D, Sunstein CR (2008) Climate change and discounting the future: a guide for the perplexed. Harvard Law School Program on Risk Regulation Research Paper No 08-12 at https://www.hks.harvard.edu/m-rcbg/cepr/Online%20Library/Papers/Weisbach_Sunstein_Climate_Future.pdf. Accessed 22 Aug 2016

White L Jr. (1967) The historic roots of our ecologic crisis. Science 155:1203–1207

Chapter 15
Human Capital in a Climate-Changed World

Shi-Ling Hsu

Abstract At the center of the crisis of climate change is an amazingly efficient but inert fossil fuel-centered energy industry. What makes the energy industry so inert is its massive stock of capital: the facilities, structures, networks, and other physical assets required to extract, process, distribute, and combust fossil fuels. This capital stock is predicated on fossil fuel exploitation, and does not adapt well to alternative methods of meeting energy needs. Fossil fuel subsidies have bloated capital investments in the energy sector, producing low prices and in turn, economic development. It has thus been widely assumed that this is the most reliable model for economic development. Two things have become increasingly clear: (i) that fossil fuel subsidies and low energy prices are not a condition precedent to economic development, and (ii) human capital development, primarily through broad provision of education, *is* a condition precedent to economic development. The climate crisis highlights the environmental harms of fossil fuel combustion, but it also shines a spotlight on the faulty economic reasoning behind a fossil fuel-centered model of economic growth. This chapter suggests that as a minimum, two no-regrets policies be linked: the phase-out of fossil fuel subsidies, and the robust financing of broader access to education. The climate crisis introduces a new set of inequalities, those in which some countries have benefited disproportionately from the combustion of fossil fuels, and a mostly different set of countries will suffer disproportionately from the harms of climate change. Effecting a direct transfer of fossil fuel subsidies to educational objectives simultaneously reduces the inert capital in fossil fuel industries, increases more productive capital in the form of human capital, and provides a compensatory mechanism for those disproportionately harmed by climate change. Ultimately, the most beneficial and lasting aid that can be provided for developing countries most vulnerable to climate change is one that increases human capital through broad educational initiatives.

S.-L. Hsu (✉)
D'Alemberte Professor of Law and Associate Dean for Environmental Programs, College of Law, Florida State University, Florida, USA
e-mail: shsu@law.fsu.edu

Introduction

Having already imposed costs on human societies likely in the hundreds of billions of dollars,[1] and with the worst still ahead,[2] it is clearly not an exaggeration to label climate change as a "crisis." In addition to the threat to humankind generally, much has been made of the fact that certain harms from climate change will accrue unevenly. Moreover, contributions to the buildup of greenhouse gases have been (and continue to be) very uneven.[3] In the light of these climate change-related welfare disparities, several redistributive proposals have been put forth to disgorge those disproportionately responsible, and for those parties to compensate those disproportionately harmed. Broadly speaking, these proposals seem to fall under the label of "climate justice." But in the calls for climate justice, the actual redistributive remedies are unclear. If there is a payment of money, how and to whom will it be disbursed? From whom and how much should a contribution be owing? Since the costs of climate change are expected to increase over time, should compensation be set aside for future generations?

This chapter focuses on one critical, mostly under-analyzed aspect of the global economy, as the key to climate justice: the role and nature of *capital*. Broadly speaking, capital is any asset that generates some future benefit or stream of benefits. Investment in capital is central to economic growth, and the enormous scale of capital in developed economies means that it inevitably plays a central role in determining the direction of economic growth. Massive investment in fossil fuel-centered methods of energy generation, transmission and consumption has meant that developed economies have unsurprisingly evolved around fossil fuels, to the exclusion of a vast array of alternatives.

The climate crisis requires us to reconsider how we think about capital and economic growth, and how government policy should treat capital. Clearly, it is a grave and monumental mistake for the world to have arrived at this ecological precipice; that it has done so in large part due to the accumulation of trillions of dollars of fossil fuel-centered capital is cause for examining the assumptions underlying this capital investment. In so doing, lessons can be learned for both mitigation policy and for adaptation policy, while putting forth the best way of compensating developing countries that have contributed little but will suffer the most. In particular, a focus on capital can help avoid making one crucial mistake all over again: making big bets on large, expensive physical capital that cannot be undone if circumstances or knowledge changes. This was the mistake that was made in building up the world's trillions of dollars of capital used to produce energy using fossil fuels, the very presence of which has delayed, perhaps catastrophically, the advent of climate policy.

---

[1] See, e.g., Fundación DARA Internacional (2012), at 17.
[2] See, e.g., Dietz et al. (2016).
[3] Caney (2014).

The Nature of Capital

Capital and labor are the two stylized inputs to production. Given the abundance of labor in developing countries, it is clear that capital is usually the limiting ingredient of economic growth.[4] Furthermore, more capital is better. Adding capital never decreases output.[5] Capital may be costly and may not be worth the cost, but capital is never modeled as having negative productive value.[6]

Despite the central role that capital plays in economic models,[7] a widely-accepted definition is lacking. Adam Smith defined capital as "[his] stock … which, he expects, is to afford him his revenue."[8] Robert Solow has defined it in passing as a "stock of produced or natural factors of production that can be expected to yield productive services for some time."[9] Gregory Mankiw defines capital as foregone current consumption to produce more income later.[10] In my earlier work I adopted a broad working definition of capital as an asset that generates a stream of benefits over time.[11] Under this broad definition, a very wide variety of equipment, structures, machines and other assets are capital that serve as an engine for economic trade, growth and prosperity. In energy industries, capital includes power plants, refineries, oil rigs, natural gas processing plants, electricity transmission and pipeline networks, and many other large, structural resources that together form an efficient energy extraction, processing, delivery, combustion, and consumption system.

Although capital can take many forms,[12] it is most easily conceived of as physical capital, such as an industrial facility that is large and expensive but produces some good in mass quantities over an extended period of production.

---

[4]Solow's fundamental neoclassical growth model posits growth as a general function of labor, capital, and technology, the latter being a multiplier that makes the other two inputs more productive. Solow (1956).

[5]Idiosyncratic exceptions may exist, but the Cobb-Douglas production function is almost never deployed with capital having an inverse relationship with productivity.

[6]The Cobb-Douglas production function, which every economics student learns about in undergraduate economics, posits production as a function of the quantity and productivity of just two types of inputs: labor and capital. Cobb and Douglas (1928). The now-familiar Cobb-Douglas formulation, $Y = AL\alpha K\beta$, with Y representing output, L representing labor, and K representing capital, has become a foundational relation in economic theory.

[7]Solow (1956), at 70.

[8]Smith (1766).

[9]Robert M. Solow, 'Notes on Social Capital and Economic Performance' in P. Dasgupta and I. Serageldin (eds.), *Social Capital: A Multifaceted Perspective* (World Bank, 2000), pp. 6–9, at 6.

[10]Mankiw et al. (1995), at 293.

[11]Hsu (2014), at 729.

[12]Financial capital is not discussed because it does not produce the kind of path-dependency problems identified in this article. Natural capital, ecosystems that produce environmental services, are also excluded because they are generally not owned, and thus are not sought to be protected by rent-preserving activities. Social capital is not discussed in this article because legal rules do not promote their formation or protect existing social capital.

The nature of such large, expensive physical capital—power plants, refineries, oil rigs, steel mills, cement factories, and other brick-and-mortar investments—is that they produce some commodity in large quantities, and only gradually, over relatively long periods of time, pay for themselves. Economies of scale tend to be important to production of these commodities, so that such machinery *must* be large and long-lived. Many of these commodities are traded globally and are sold at fairly thin profit margins, so that competitiveness is often important to the owners of this capital.

The problem with this large-capital, low-cost model of production is that large capital is vulnerable to changes in the legal and economic environment.[13] At any given time, capital may be rendered obsolete by regulatory changes or changes in the economic environment. The value of the typically very high-cost capital in oil extraction and refining have been dramatically reduced by a nearly two-year decline in the global market price of oil.[14] Climate policy, if effective, could also severely reduce the value of this capital. It is one thing to lose value due to market conditions, but it is another to lose it due to regulation; the latter can be resisted. Given the large stakes in perpetuating operation of expensive capital, the owners of this capital become very protective, seeking to ensure that the legal and economic environment in which they operate remain static long enough for them to realize some reasonable return. When threatened with some change in its economic environment, owners of capital will naturally attempt to influence law and policy so as to preserve the value of their capital.[15] What we observe under these circumstances are *rent-preserving activities,* the ex post analog of rent-seeking, and the exercise of protecting existing rents.[16] As opposed to attempts to affirmatively procure legal privilege that is privately advantageous but publicly costly, rent-preserving activities are attempts to preserve existing privilege.[17] Although capital is generally an economic good, the little-appreciated downside is that once it is in place, it creates its own political economy to protect it. Given the vast size of capital stocks, they hold the potential to dramatically change the political economy of regulation, trade, and other changes in the economic and legal environment.

Capital can also be *human* capital, the formal and informal education and on-the-job training that enable people to perform skilled productive tasks.[18] Like physical capital, human capital generates a stream of benefits, and is a powerful ingredient for economic growth. Like physical capital in the fossil fuel-centered industries, it can create its own political economy against reform. However, human

---

[13] Hsu, 11, at 735–43.

[14] 'BHP Billiton Takes £5bn Writedown on US Oil Assets as Price Slump Takes Toll' (2016) *The Guardian,* January 14, 2016, available at: https://www.theguardian.com/business/2016/jan/15/bhp-billiton-takes-5bn-writedown-on-us-oil-assets-as-price-slump-takes-toll.

[15] Olson (1982), at 41–47.

[16] Olson (1982), at 41–47.

[17] See, e.g., Baumol and Ordover (1985).

[18] Becker (1993), at 30–54.

capital can be a much more flexible form of capital, so that it might be deployed for a number of different purposes. Human capital is simply useful knowledge, and can thus be narrow, specific knowledge—in which case could give rise to rent-preserving activities—or it can be broad, general knowledge, which might be less vulnerable to changes in the legal and economic environment. Moreover, human capital has been a more reliable stimulant of economic growth than the physical capital embedded in fossil fuel-centered energy industries. Along these lines, this chapter proposes that investment in human capital is likely to be a better economic development strategy than using fossil fuels to ensure low energy prices.

How Capital Has Stalled Climate Policy

Energy capital has accomplished, on a grand scale, what economists hope for: economic development. Low energy prices produced by fossil fuel exploitation have enabled an enormous amount and range of economic activity, giving rise to tremendous economic growth over the last century. The positive externalities are very large. But the negative externalities have also been very large, and the fossil fuel-centered energy industries have done their best to obscure them.

Fossil fuel-related industries are among the most capital-intensive industries in the world.[19] Changes in environmental and other regulations can severely affect the profitability and therefore the value of fossil fuel-related capital, and are vigorously resisted. It is in this challenging political environment that climate policy has operated: fossil fuel-centered energy industries with capital that is expected to generate benefits for a long period of time, find their capital threatened by climate policies that impose additional, potentially crippling costs on fossil fuel-related operations. Resistance to this policy threat has included an extensive public relations campaign to sow doubt about the existence and seriousness of climate change.[20]

Historically, the compensating benefit of fossil fuel-centered energy industries has been the predictability of revenues, as developed economies have always been highly dependent upon energy as a staple of economic growth. Fossil fuel-centered industries could also, until recently, look forward to the promise of expanding production and sales in developing economies, as economic development was thought to bring demand for fossil fuel-derived energy. The massive scale of production and consumption generates large profits, and is itself a powerful lure for investment. Upsetting this paradigm is thus upsetting a very large apple cart. Publicly-traded energy companies in Canada and the U.S. made profits of

---

[19]See, e.g., Chow et al. (2003), at 1529.
[20]See, e.g., Oreskes and Conway (2011).

$257 billion in 2014, which is greater than the GDP of Chile.[21] That figure excludes the privately-held Koch Industries, which itself had estimated revenues (not profits) of $100 billion.[22]

To protect this mode of business, fossil fuel-centered industries have demonstrated their ability and their inclination and ability to engage in rent-preserving activities. A push in the 1990s and early 2000s to deregulate retail electricity markets in the United States has foundered.[23] Initially, deregulation was favored by nearly all stakeholders, from integrated electric utilities to consumer groups to rural electric cooperatives. But when proposals became concrete and winners and losers became tangible, opposition hardened.[24] In particular, utilities worried about what would happen to their "stranded assets," their capital that would be rendered uncompetitive by a new, deregulated, and more competitive electricity marketplace.[25] The stranded assets that took center stage in the deregulation debate were mostly in the form of coal-fired power plants. A deregulated electricity generation environment was expected to render many of those coal-fired power plants obsolete.[26] In some unexpected ways, this has come to pass, at least in jurisdictions in which competition for the retail provision of electricity is allowed.[27] This trend has been amplified by the advent of hydraulic fracturing, which has dramatically lowered natural gas prices so as to render coal uncompetitive as a fuel source.[28]

In those jurisdictions that have not taken up electricity deregulation, and remain in a regulated utility legal regime,[29] fossil fuel-centered electric utilities have continued to fend off innovation and evolution of their industry.[30] Non-utility firms now exist to install solar panels on the roofs of private residences, allowing the

---

[21]Shakuntala Makjijani and Lorne Stockman, *Despite Falling Prices North America's Fossil Fuel Sector Makes Healthy Profits*, Oil Change International, May 5 2015; online: http://priceofoil.org/2015/05/05/despite-falling-prices-north-americas-fossil-fuel-sector-makes-healthy-profits/.

[22]Murphy (2016).

[23]The U.S. Energy Information Administration considers fifteen states as "active" in deregulation or "restructuring," and seven in a "suspended" mode of deregulation. U.S. Energy Information Administration, Status of Electricity Restructuring by State, http://www.eia.gov/electricity/policies/restructuring/restructure_elect.html (September 2010). Other definitions of "deregulation" may yield different results. See, e.g. Severin Borenstein and James Bushnell, 'The U.S. Electricity Industry After 20 Years of Restructuring', (2016) __ *Annual Review of Economics*, at _ [7–8]_ (forthcoming), available at: http://papers.ssrn.com/sol3/papers.cfm?abstract_id=2640081.

[24]Cearley and Cole (2003).

[25]Brennan and Boyd (1997), at 42.

[26]Brennan and Boyd, n. 20 above, at 42.

[27]Power plants in states that deregulated electricity generation employed about 6% fewer employees after deregulation and, incidentally, enjoyed a 13% decrease in nonfuel operating expenses. Fabrizio et al. (2007), at 1266–69 (Tables 4 and 5).

[28]See, e.g., Merrill and Schizer (2013), at 148.

[29]For a review, see Gilbert et al. (1996), at 2–3.

[30]Sine and David (2003), at 193.

homeowners to defray the cost of utility-provided electricity,[31] but face resistance. In the U.S. State of Florida, which enjoys strong solar energy resources,[32] utilities have aggressively resisted efforts to liberalize entry. Florida law allows only regulated utilities to sell electricity in the retail market,[33] which is an important prohibition, because if an individual homeowner sought to finance her residential rooftop system, the financier may be considered a "utility" that would have to comply with all of Florida electric utility regulations.[34] Complying with the quagmire of utility regulations is a prohibitive cost to non-utility firms, which lack the resources and legal expertise that regulated utilities possess. The sizable compliance department needed to operate as a regulated utility is generally only feasible for large regulated utilities, which can spread the costs over their captive ratepayers, a luxury start-up companies with creative energy ideas cannot afford. A ballot initiative to liberalize electricity generation by relaxing the scope of regulation and facilitate the installation of rooftop solar photovoltaic panels was met with a strong industry-led campaign and a *competing* ballot initiative to prevent potential competitors from making inroads on the customers of regulated utilities.[35]

In this context, it should not be surprising that fossil fuel-centered energy industries should resort to litigation,[36] political influence,[37] and even manipulating public opinion[38] to forcefully contest policies that threaten their capital. But the exact nature of their interest has not been carefully scrutinized. It turns out that their engine of growth and efficiency—their capital—is the source of vulnerability, and the impetus for zealously protecting its own fragile economic and legal environment.

---

[31]Cardwell (2015), *New York Times*, October 2, 2015, at B2, available at: http://tinyurl.com/omvq2q6.

[32]National Renewable Energy Laboratories, Solar Maps (2015), available at: http://www.nrel.gov/gis/solar.html (accessed December 14, 2015).

[33]Florida Statutes § 366.82(1)(a).

[34]*PW Ventures v. Nichols*, 533 So. 2d 281 (1988).

[35]Jim Turner, 'Solar Choice Ballot Initiative Targets 2018' (2016) *Sun Sentinel*, March 6, 2016, available at: http://www.sun-sentinel.com/business/consumer/fl-nsf-solar-choice-2018-ballot-20160111-story.html.

[36]See, e.g., *In re Murray Energy Corp.*, 788 F.3d 330 (D.C. Cir. 2015); *cert. granted*, 136 S. Ct. 999 (2016).

[37]Newell and Paterson (1998).

[38]See, e.g., Gillis and Krauss (2015); Krauss (2007), at C7; online: http://www.nytimes.com/2007/01/04/business/04exxon.html?_r=0.

Rigid Capital

Capital impedes legal reform because new regulation is costly to the capital owner. This may be the case when the impetus for reform is the identification of negative environmental externalities, and production must add on some pollution-reducing equipment, or shift entirely to a new mode. The problem arises when capital that is deployed for a specific production method cannot be easily re-deployed to a different, less environmentally harmful production method. If capital is specific enough, then any legally-mandated shift in production methods could effectively "strand" that capital and render it worthless.

But capital *could* be less of an obstacle if it were flexible, and susceptible of redeployment. After all, if an asset can be used for a variety of purposes, in a variety of industries, then it will continue to hold value even if the regulatory environment or competitive conditions change. In and of itself, capital is not necessarily an obstacle to reform; it is only the *rigidity* of capital that should raise "what-if" questions about the possibility of future regulation or obsolescence.

Capital in fossil fuel-related industries, unfortunately, does not tend to be flexible.[39] Fossil fuel-centered energy industries have steadily evolved for over a century so that capital embedded in the various stages of energy production are the result of continuing experimentation, development, and very large-scale production, not to mention massive government subsidization.[40] This continuing development has produced countless small operational efficiencies, which have cumulated to create increasingly efficient but task-specific capital, focused as it is on extracting ever more fossil fuels from less favorable conditions, and producing it at ever-decreasing cost. The relentless quest for technical efficiency has created tightly integrated and interdependent systems of extraction, production, transportation and distribution, delivering enormous amounts of energy at low consumer cost. The price of such technical efficiency has been flexibility.

Offshore oil rigs are an example of the extensive development and massive scale of production embodied in fossil-centered energy capital. Typical large modern offshore oil rigs, which can extract crude oil from ever-greater depths, have the capacity to extract up to 200,000 barrels of oil per day[41] over their typically thirty-year lifetimes.[42] A current state-of-art cost rig carries a price tag in the billions of dollars.[43] Profit is highly dependent upon economic conditions (as the recent global plunge in crude oil prices demonstrates), and is highly dependent

---

[39]Joskow (1991), at 67.

[40]Coady et al. (2015).

[41]See, e.g., *Atlantis Deepwater Oil and Gas Platform Gulf of Mexico, United States of America*, OffshoreTechnology.com (no date), online: http://www.offshore-technology.com/projects/atlantisplatform/.

[42]Chevron, *Jack/St. Malo* (2001), online: https://www.chevron.com/projects/jack-stmalo.

[43]*Id.*

upon a stable regulatory environment. Most importantly, an offshore oil rig can be used for only one purpose: drilling for oil in oceanic waters.

Coal-fired power plants have also served the cause of economic development, delivering large amounts of electricity at low consumer cost.[44] But these large, multi-billion dollar plants, with their enormous boiler chambers, turbines, and smokestacks and even pollution control equipment are designed to carry out one function in one way: burn large amounts of coal to spin a very large turbine to generate an electrical current. These coal-fired power plants can be converted into natural gas-powered power plants (which emit much less pollution),[45] but the process is expensive.[46] Coal-fired power plants are thus another example of the rigidity of expensive capital in fossil fuel-centered energy industries.

If economic conditions or environmental regulations put a halt to operations, the owner of those oil rigs and power plants will suffer huge losses. In a post-carbon world, there is little or no salvage value for offshore oil rigs, coal-fired power plants, or any number of large expensive, single-purpose facilities predicated on the combustion of fossil fuels. It is as if the fossil fuel-related industries have, like complex and undisturbed ecosystems, created narrow ecological niches for many species of physical and human capital. The disadvantage of such finely-tuned and interdependent systems is that they have evolved into highly specific and interdependent parts, and are vulnerable to disturbance. The response has been predictable: these industries have not permitted their capital to fail. American trade groups such as the American Petroleum Institute and the Edison Electric Institute have exercised enormous power over the U.S. legislative and administrative processes,[47] as has the Canadian Association of Petroleum Producers over Canadian government.[48] Governments heavily dependent upon oil revenues, such as Saudi Arabia and Venezuela, have stalled climate policy in international negotiations.[49]

---

[44]That is to say, purchasing electricity has been inexpensive, even if the external costs have been large. A recent study of the net economic benefits of a wide variety of industries in the U.S. found that coal-fired electricity production almost certainly subtracts more from GDP in the form of health and environmental harms than it contributes, in the form of electricity provision. Muller et al. (2012).

[45]U.S. Energy Information Administration, Frequently Asked Questions, 'How Much Carbon Dioxide is Produced When Different Fuels Are Burned?' available at: https://www.eia.gov/tools/faqs/faq.cfm?id=73&t=11.

[46]Scott Gossard, 'Coal-to-Gas Plant Conversions in the U.S.' (2015) *Power Engineering*, June 18, 2015, available at: http://www.power-eng.com/articles/print/volume-119/issue-6/features/coal-to-gas-plant-conversions-in-the-u-s.html.

[47]Darren Samuelson & Katherine Ling, 'Fragile Compromise of Power Plant CEOs in Doubt as Senate Debate Approaches' *E&E News*, 5 August 2009.

[48]Michael Bolen, 'Peter Mansbridge Was Paid by Oil and Gas Lobby for Speech' *The Huffington Post Canada*, 26 February 2014, available at: http://www.huffingtonpost.ca/2014/02/26/peter-mansbridge-oil-speech_n_4861979.html.

[49]Pew Center for Global Climate Change, 'Fifteenth Session of the Conference of the Parties to the United Nations Framework Convention on Climate Change and the Fifth Session of the Meeting

Better Capital: Human Capital

It is indisputably true that low energy costs have spurred economic development. Developing countries have only been left out because of a lack of money needed to acquire energy capital. But apart from the aggressive industrial interventions in policy to stall climate policy, the massive government subsidies for fossil fuel-related industries, under the guise of capital formation for economic development, call into question the intrinsic value of fossil fuels. A recent study by the International Monetary Fund estimated that eliminating energy subsidies would raise government revenue by $2.9 trillion, reduce global CO_2 emissions by 20%, and halve premature air pollution deaths.[50] Achieving low energy prices, powered by fossil fuels, has long been the dominant paradigm of economic development. As it happens, old paradigms of economic development and capital investment fade slowly, in large part due to the sheer size and ubiquity of embedded capital.

The climate crisis provides the impetus and an opportunity to cast off anachronistic mindsets on capital and economic development. If there is a public purpose of capital, it is to generate economic growth. The problem with industry captains is that they have conflated the two; their capital has become their own *raison d'etre*. Large financial exposure due to the high costs of capital have led fossil fuel-centered industries to distort public policy to protect their capital, even if economic growth can be decoupled from fossil fuel-centered energy production. It seems quite likely that titans in fossil fuel-centered industries, faced with mounting evidence of the massive (and growing) environmental harms of their operations, have responded by deluding themselves into believing that low fossil-based energy prices are still the key to economic salvation.[51]

At the same time, international relations continue to play a central role in both domestic and international climate policy, which have implications for capital. Developing countries strongly reject any suggestion that they be deprived of the same economic opportunities already enjoyed by developed countries. But development in this much-larger segment of the world's population clearly cannot be achieved by the formation of the same kind of energy capital as that which fueled economic growth in the twentieth century. Stimulating economic growth in developing countries will require new thinking on capital.

Odd as it may seem, the climate crisis is an appropriate time to reconsider the role of *human capital* in economic growth. Human capital is indisputably a critical

---

of the Parties to the Kyoto Protocol,' (2009), available at: http://www.c2es.org/docUploads/copenhagen-cop15-summary.pdf.

[50]Coady, n. 35, at 7.

[51]Examples abound, but the remarks of Robert Murray, the founder and CEO of Murray Energy, are exemplary. See, e.g., Robert E. Murray, 'Murray Energy's Strategy for Succeeding in Transitional Coal Markets' (2014), available at: http://www.eenews.net/assets/2014/09/23/document_gw_01.pdf ("The insane, regal Administration of King Obama has ignored science, economics, our poorer citizens and those on fixed incomes, our manufacturers, and the Constitution, as it has by-passed our Congress.").

driver for economic growth, and represents a much better investment than the large, expensive capital in the fossil fuel-centered industries. The climate crisis simultaneously forces developed societies to question the intrinsic value of fossil fuel-centered capital, and to cast about for a better path to economic prosperity, particularly for developing countries.

Economists have long recognized the central importance of human capital to economic growth.[52] Nobel Laureate Robert Lucas's seminal comparison of South Korea and the Philippines from 1960 to 1987 attributed the "miracle" of rapid South Korean economic growth to government policy focusing on broad and rigorous public education.[53] To illustrate Lucas's point, Fig. 15.1 below shows GDP levels and percentage of population with tertiary education (as a measure of human capital) in the two countries from 1950 to 2010.

While Lucas elides some obvious and important historical and cultural differences, his study continues to stand as a seminal work on the importance of education as the development of a stock of human capital. More recently, economists Claudia Goldin and Lawrence Katz, in their book *The Race Between Education and Technology*,[54] argue that the economic dominance of the United States for the latter half of the twentieth century was largely due to its broad and compulsory public schooling system, which created an educated work force able to adapt to technological changes and increase productivity.[55] By contrast, American underperformance since 1970 relative to other countries is, argue Goldin and Katz, largely attributable to the American failure to maintain that educational advantage.[56]

Despite consensus among economists of the value of human capital, it remains under-supplied for a number of reasons. First, all other things being equal, human capital is a riskier investment for an individual than an investment in physical capital. That is say, if an expected return on physical capital such as a hot dog stand is equal to the expected return on human capital such as a bachelor's degree in English, a risk-averse individual would be more inclined to invest in the hot dog stand. That is because human capital cannot be bought or sold like physical capital can, so diversifying a capital stock requires more time and resources normally available to an individual.[57] By contrast, the transferability of physical capital means that an individual does not need to diversify; a diversified economy does this.[58] In light of this individual bias towards physical capital, a capital policy should generally be more generous towards human capital than physical capital.

---

[52]See, e.g., Schultz (1960); Theodore W. Schultz, *Investment in Human Capital*, 51 A. Econ. Rev. 1 (1961).

[53]Lucas, Jr. (1993).

[54]Claudia Goldin and Lawrence F. Katz, The race between education and technology (Harvard Univ. Press, 2008).

[55]Goldin & Katz, *supra*, note 50, at 29.

[56]Goldin and Katz, *supra*, note 50, at 320–23.

[57]Levhari and Weiss (1974).

[58]*Supra*, note 53, at 950.

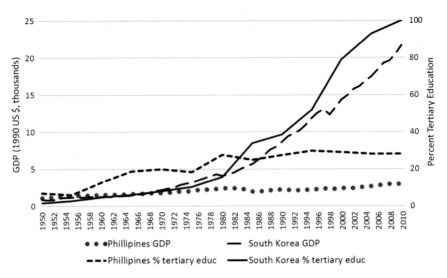

Fig. 1 GDP and percent tertiary education

Secondly, human capital confers positive externalities in a way that physical capital generally does not. Human capital is knowledge, and the greater the stock of human capital, the greater the knowledge spillovers, and the higher the rate of accumulation of human capital. Knowledge begets more knowledge, and does so more easily if there is more knowledge to begin with. The formation of human capital should thus be made with a view toward taking advantage of positive consumption externalities. Estimates of the value of human capital bear this out: the value of human capital in the United States is in the neighborhood of 11 to 16 times the value of physical capital, but investment in human capital is only about four times that of physical capital.[59]

More Flexible Capital

It might seem odd to compare public investment in human capital with private investment in fossil fuel-centered capital. But in addition to being more effective in stimulating economic growth, human capital has one other critical advantage that happens to highlight, in contrast, the folly of fossil fuel capital: flexibility. Human capital is generally, by its nature, susceptible of multiple uses. Clearly, human capital could be specific—the knowledge of how to accomplish very specific tasks —or human capital could be general—broad understandings of some principles, without necessarily knowing how to accomplish specific tasks. Specific human

---

[59]Jorgenson and Barbara (1989), at 228; Christian (2010), at 34.

capital, however, is built upon a foundation of general human capital. The ability to perform specific tasks is the ability to apply knowledge, which presupposes the more general form of knowledge. Basic principles of physics give rise to the ability to apply them to a wide variety of engineering disciplines; basic principles of biology give rise to the ability to apply them to a wide variety of life sciences, including agriculture. Most fundamentally, reading is a basic skill upon which almost all other forms of knowledge derive. Development of human capital is thus inherently development of a more flexible asset than the kind of expensive physical capital that has been invested in fossil fuel-centered industries.

Human capital in the fossil fuel-centered industries *could* be flexible, as other technical, engineering-heavy industries are. But here again, fossil fuel-centered industries have demonstrated no interest in making their labor force more flexible.[60] Workers on offshore oil rigs, for example, require no formal education; unskilled workers are required to take training courses certified by a trade association called the International Well Control Forum. After completion, "drillers," "rig mechanics," "subsea engineers," and "derrickmen" are among the crew of 80 workers that can earn $50,000–$100,000 for six months' work.[61] But these workers are not equipped (at least not by their oil rig training) to do anything else. Most of the tens of thousands of laid-off workers in the struggling Canadian oil sands industry have not found re-employment.[62] Engineering principles learned by petroleum or power engineers *could* be portable, but fossil fuel-centered industries have eschewed more general training. In part because of the public good nature of human capital, fossil fuel-related industries have financed only that capital that serves their specific production needs, leaving the broader, more general educational tasks to formal schooling.[63] The narrowness of even the human capital in fossil fuel-centered industries is all the more reason to completely rethink the nature of capital.

It is difficult to predict what the most useful form of capital will be just a few years in the future. But the more *general* the capital, the more flexible it will be, and the more likely it could be redeployed should a product or process become obsolete. The mistake in developing through exploitation of fossil fuel sources was that the embedded capital was inflexible, and served to lock out alternative systems. For that matter, much international aid intended to relieve poverty in underdeveloped

---

[60] See, e.g., *Skills Needs in the Energy Industry* (Energy Institute, 2008), available at: https://www.energyinst.org/documents/5.

[61] See, e.g., Claire Calkin, 'Offshore Oil Rig Jobs Can Be Tough, But Very Rewarding' (no date), available at: https://www.experience.com/alumnus/article?channel_id=energy_utilities&source_page=additional_articles&article_id=article_1128902416846.

[62] Dawson (2015).

[63] *Human Capital Strategies for Canada's Energy Sector* (Mercer, 2010), at 4, available at: https://www.conference-board.org/retrievefile.cfm?filename=Human-Capital-Strategy-for-Canadas-Energy-Sector.pdf&type=subsite.

countries have been mis-spent on dubious capital projects, including dams, irrigation systems that ultimately prove culturally or geographically inappropriate.[64]

A Proposal

The climate crisis thus offers not only the opportunity, but the impetus to revisit the nature of development strategies, and of aid to countries that are likely to suffer disproportionately in a climate-changed world. It has certainly been argued that intergenerational justice does not *necessarily* militate in favor of minimizing risk, but of some combination of minimizing risk and enhancing resilience. In that light, low energy prices might not be such an inefficient state of affairs if economic prosperity can fund resilience.[65] But given the severity and immediacy of the fossil fuel-induced climate crisis, and given the readiness and superiority of an alternative development strategy—human capital development—continuing down a path of fossil fuel-centered economic growth seems foolhardy.

At least one simple link can be made between reform of fossil fuel-centered capital and a human capital strategy for development: at a minimum, the massive government subsidies for fossil fuel-centered capital—about $3 trillion, not counting the uncompensated environmental costs[66]—would unquestionably be better spent in enhancing public education. Even if developed countries suffer what is perceived as distortions and pathologies in public education funding,[67] it pales in comparison with the distortionary effects of fossil fuel subsidization. Removal of energy subsidies has been estimated to produce a $1.8 trillion increase in global wealth, even fully accounting for the welfare effects of higher energy prices. No serious argument can be mounted against this $3 trillion being better spent on public education. Moreover, no serious argument is made anywhere that government spends *too much* money on public education.[68] Outlays for subsidies for fossil fuels could be redirected to public education, a huge boon worldwide.

Moreover, any redistribution in the form of education should include a preference for human capital for younger people. Skills beget skills and learning begets

---

[64]See, e.g., Williamson (2010), Boone (1996), Lavagnon (2012), Clark (2002), Anne Danaiya Usher, Routledge Studies in Development and Society: DAMS as AID (1) 5 (A. Usher, ed., Routledge, 2005).

[65]Cox and Cox (2016).

[66]Coady et al., n. 35, at 7.

[67]See, e.g., Lipman (2004); Mark Gradstein, *The Political Economy of Public Spending on Education, Inequality, and Growth* (World Bank Policy Research Working Paper No. 3162, 2003), *available at* http://elibrary.worldbank.org/doi/pdf/10.1596/1813-9450-3162.

[68]For a widely praised book on the role of education in economic growth, see, Claudia Goldin and Lawrence Katz, The Race Between Education and Technology (2008). See also, Lionel Artige & Laurent Cavenaile, *Public Education Expenditures, Growth and Income Inequality* (2016) available at: http://papers.ssrn.com/sol3/papers.cfm?abstract_id=2759093.

learning, so that the earlier in life an individual acquires skills (i.e., forms human capital), the more likely she is to build on those skills to acquire other skills.[69] If resources are scarce, then the best investment of government dollars for human capital promotion would target young children for early childhood learning and social skill development.[70]

But even beyond the modest prescription of a one-for-one substitution in spending—public education for fossil fuel subsidies—it is worth considering the role of human capital in a broader context, and against the backdrop of climate justice. If developing countries clamor for an equalization of opportunity and an uplift in their standard of living, by far the best way to accomplish that is to directly fund public educational systems and not, as anachronistic development models would posit, low energy prices. Too often, foreign aid programs have failed to alleviate poverty because dollars spent in economic development have found their way into dubious ventures, and boosted economies only artificially, failing to improve the fundamentals of a populace.[71]

The climate crisis also provides an opportunity to rethink the nature of justice. To be sure, adaptive capacity must be a large part of relief for underdeveloped countries that will suffer disproportionate harms from climate change. And to be sure, developed countries benefiting from the fossil fuel-led economic growth should take the lead in funding and finding the technological breakthroughs necessary to mitigate and adapt to climate change, and possibly even "geo-engineer" artificial changes to earth systems to reverse emissions by re-capturing and sequestering greenhouse gases already emitted. But a long and tortured history of largely unsuccessful uses of international aid to alleviate poverty and stir economic development provides a cautionary tale in terms of how to provide effective compensation for those suffering disproportionately from climate change.

Throughout economic history, only one use of money has proven to be reliable in generating economic growth. Ill-advised and culturally or geographically ignorant capital projects aimed at economic development have frequently failed to deliver on the promise of economic development, and have mostly left just different miseries in their wake.[72] Nothing has ever worked as well in lifting up a populace as broadly educating it.

---

[69] James J. Heckman, 'Policies to Foster Human Capital' (2000) 54 *Research Economics* 3–56, at 5 ("Early learning begets later learning…").
[70] Heckman, *supra*, note 63, at 5–6.
[71] *Supra*, note 60.
[72] *Supra*, note 60.

Conclusion

A lawyer's call for "climate justice" is apt to include some demand for compensatory payment[73] or, on the fringes, injunctive relief for some greater governmental climate action.[74] While this could offer some visceral satisfaction in the unlikely case of success, it is not clear that recovering plaintiffs are ultimately much better off. It is as if, for lawyers, justice stops when a transfer of wealth is made; what happens after that is presumably left to others.

Such blitheness is likely to repeat the failures that have plagued previous international efforts to alleviate poverty in underdeveloped countries. Since there is a great deal of overlap between countries receiving foreign aid and those likely to suffer disproportionately from climate change, the climate crisis is an opportunity an opportunity to rethink the nature of economic development, and start anew. Effective compensation must take the form of capital – otherwise compensation is never sustaining —but it must take a particular form of capital. Compensatory capital must generate economic growth, and must be flexible. A human capital approach offers the best approach to leveling the benefits and burdens of climate change.

It has long been obvious that fossil fuel subsidies exacerbate pollution and reduce welfare. Of all of the negative externalities of a fossil fuel-centered energy industry, the emission of greenhouse gases is only the latest one. However, the severity and immediacy of the climate crisis can be the impetus for a re-imagining of the capital needed to sustain economic development. The climate crisis is a context in which a direct comparison of fossil fuel subsidization and human capital development becomes a logical one. Both are economic development strategies, and exchanging the former for the latter is also a compensatory mechanism for those countries that are likely to be disproportionately harmed by climate change. Certainly, much good could be done with the roughly $3 trillion in global fossil fuel subsidies, but as a climate policy, such a direct transfer from one type of capital to another becomes a coherent, self-contained policy.

Acknowlegement The author would like to thank Brian Labus and Christopher O'Brien for their research assistance, and the always helpful Florida State Law Library staff.

---

[73]See, e.g., Farber (2007).

[74]John Schwartz, 'In Novel Tactic on Climate Change, Citizens Sue Their Governments' (2016) *New York Times,* May 12, 2016, at A6, available at: http://www.nytimes.com/2016/05/11/science/climate-change-citizen-lawsuits.html?emc=edit_th_20160512&nl=todaysheadlines&nlid=66362416&_r=0.

References

Baumol WJ, Ordover JA (1985) Use of Antitrust to Subvert Competition. J L Econ 28:247–249
Becker GS (1993) A theoretical and empirical analysis, with special reference to education, 3d edn
Boone P (1996) Politics and the Effectiveness of Foreign Aid. Eur Econ Rev 40:289
Brennan TJ, Boyd J (1997) Stranded costs, takings, and the law and economics of implicit contracts. J Reg Econ 11:41–54
Caney S (2014) Two kinds of climate justice: avoiding harm and sharing burdens. J Political Philoso 22:125–149
Cardwell D (2015) Solarcity to make high-efficiency panel. New York Times
Cearley RW, Cole DH (2003) Stranded benefits versus stranded costs in utility deregulation. In Grossman PZ, Cole DH (eds) The economics of legal relationships: the end of a natural monopoly: deregulation and competition in the electric power industry
Chow J, Kopp RJ, Portney PR (2003) Energy resources and global development. Science 302:1528–1531
Christian MS (2010) Human capital accounting in the United States, 1994–2006. Survey of Current Business
Clark DL (2002) The world bank and human rights: the need for greater accountability. Hum Rts J 205:211–219
Coady DP, Parry I, Sears L, Shang B (2015) How large are global subsidies? In: International monetary fund working paper 15/105, available at: http://papers.ssrn.com/sol3/papers.cfm?abstract_id=2613304
Cobb CW, Douglas PH (1928) A theory of production, 18 (supplement) Am Econ Rev 139
Cox LA, Cox ED (2016) Intergenerational justice in protective and resilience investments with uncertain future preferences and resources Risk Gov Soc 19:173
Dawson C (2015) Canadian oil-sands producers struggle. Wall Street J
Dietz S, Bowen A, Dixon C, Gradwell P (2016) Climate value at risk of global financial assets. Nat Climate Change online: http://www.nature.com/nclimate/journal/vaop/ncurrent/pdf/nclimate2972.pdf
Fabrizio K, Rose NL, Wolfram C (2007) Does competition reduce costs? Assessing the impact of regulatory restructuring on U.S. Electric generation efficiency. Am Econo Rev 97:1250–1277
Farber DA (2007) Basic compensation for victims of climate change. Uni of Pennsylvania Law Rev 155:1605–1656
Fundación DARA Internacional (2012) Climate vulnerability monitor 2d edn
Gilbert RJ, Kahn EP, Newbery D (1996) Introduction: international comparisons of electricity regulation. In Gilbert RJ, Kahn EP (eds) International comparisons of electricity regulation. Cambridge University Press, pp 1–24
Gillis J, Krauss C (2015) Exxon mobil investigated for possible climate change lies by New York Attorney General, N.Y. Times, November 6, 2015, a online: http://www.nytimes.com/2015/11/06/science/exxon-mobil-under-investigation-in-new-york-over-climate-statements.html?_r=0
Hsu S-L (2014) Capital rigidities, latent externalities. Houston Law Rev 51:719–779
Jorgenson D, Barbara MF (1989) The accumulation of human and nonhuman capital, 1948–84. In: Lipsey RE, Tice HS (eds) The measurement of saving, investment, and wealth. University of Chicago Press
Joskow PL (1991) The role of transaction cost economics in antitrust and public utility regulatory policies. J Law Econo Organ 75:3–83
Krauss C (2007) Exxon accused of trying to mislead public. N.Y. Times
Lavagnon A (2012) Ika, Amadou Diallo, and Denis Thuillier, Critical Success Factors for World Bank Projects: An Empirical Investigation. Int J Proj Mgmt 105:110–114
Levhari D, Weiss Y (1974) The effect of risk on the investment in human capital. Am Econ Rev 64:950
Lipman P (2004) High stakes education: inequity, globalization and urban school reform. In: Mark Gradstein, The political economy of public spending on education, inequality, and

growth (World Bank Policy Research Working Paper No. 3162, 2003), available at http://elibrary.worldbank.org/doi/pdf/10.1596/1813-9450-3162

Lucas RE, Jr (1993) Making a miracle. Econometrica 61:251–272

Mankiw G, Phelps ES, Romer PM (1995) The growth of nations. Brookings Papers on Economic Activity, pp 275–326

Merrill TW, Schizer DM (2013) The shale oil and gas revolution, hydraulic fracturing and water contamination: a regulatory strategy Minnesota Law Rev 98:145–263

Muller NZ, Mendelson R, Nordhaus W (2012) Environmental accounting for pollution in the United States economy. Am Econ Rev 101:1649

Murphy A (2016) America's largest private companies, Forbes. July 20, 2016; online: http://www.forbes.com/pictures/eggh45ejji/2-koch-industries/#3bba20474ef1

Newell P, Paterson M (1998) A climate for business: global warming, the state and capital. Int Rev Int Political Econo 5:679–703

Olson M (1982) The rise and decline of nations. Yale University Press pp 41–47

Oreskes N, Conway EM (2011) Merchants of doubt: how a handful of scientists obscured the truth on issues from tobacco smoke to global warming

Schultz TW (1960) Capital formation by education. J Polit Econ 51:571

Sine WD, David RJ (2003) Environmental jolts, institutional change, and the creation of entrepreneurial opportunity in the us electric power industry. Res Policy 32:185–207

Smith A (1766) An Inquiry into the nature and Causes of the Wealth of Nations. p 351

Solow RM (1956) A contribution to the theory of economic growth. Q J Econo 70:65–89

Williamson CR (2010) Exploring the failure of foreign aid: the role of incentives and information 23, Rev Austrian Econ 17

Chapter 16
A Wild Solution for Climate Change

Thomas E. Lovejoy

Abstract In addition to the physical impacts of climate change—the retreat of ice in most places, change in fire regimes, extreme weather events (droughts, major storms), sea level rise and ocean acidification, there are multiple biological impacts. The latter are no longer just modest changes in phenology and geographical distribution. The shift to accelerating change makes a strong case for limiting climate change to no more than 1.5° above pre-industrial temperature. That challenging goal can only be achieved by lowering greenhouse gas concentrations. Restoration of extensive historically degraded and destroyed ecosystems has the potential to substantially lower atmospheric CO_2 concentrations—hence a "wild solution" to climate change.

Introduction

In 1987—almost 30 years ago—(as part of a group gathered to advise on the future Convention on Biological Diversity), I have been told I said to then Executive Director of the United Nations Environment Program (UNEP), Mostafa Tolba, words to the effect that if we didn't address climate change we could forget about biodiversity. So what does the relationship between biodiversity and climate change look like today?

When Swedish Scientist Svante Arrhenius wrote his famous paper demonstrating the greenhouse effect (Arrhenius 1896) he was trying to answer the question "Why is the Earth a habitable temperature for humans and other forms of life. Why isn't it too cold?" The answer of course was the "greenhouse effect": the ability of the carbon dioxide and other greenhouse gases to trap radiant heat instead of it being lost to outer space. So the basic science underpinning climate change is quite venerable.

T. E. Lovejoy (✉)
George Mason University, Fairfax, USA
e-mail: tlovejoy@unfoundation.org

Not only was he able to demonstrate that the pre-industrial level of greenhouse gases (principally CO_2) was responsible for the Earth's climate and temperature, but also—remarkably with only pencil and paper—he was able to calculate what doubling pre-industrial levels of CO_2 would do to the Earth's average temperature. His calculations came out remarkably close to what modern super computer models now predict.

What Arrhenius was not aware of is the actual fluctuation in the planet's temperature in the preceding hundreds of thousands of years and, in particular, the last ten thousand years of climate stability. That period includes all recorded human history, the origins of agriculture and of human settlements. What is especially striking is that all recorded human history has occurred during this period of climate stability, and in general our plans for the future are based on the assumption of a stable climate. During that same period all ecosystems were adapting to a stable climate.

Throughout the history of life, there has been a dynamic between the living part of the planet and the climate system. By burning the energy from ancient photosynthesis trapped in the fossil fuels humans are dramatically changing that dynamic. Fortunately in that very dynamic between the living and physical parts of the planet is the potential to harness it and reduce the amount and thus challenge of climate change: essentially a "wild solution" to climate change.

Physical Impacts of Climate Change

Climate is now clearly changing. The planet is now approximately 1.0 °C warmer on average and CO_2 concentrations are in excess of 400 parts per million (vs. pre-industrial of 280 ppm).

There are already clear responses in physical nature, mostly around the solid and liquid states of water. In the Northern Hemisphere, lakes are freezing later in the autumn and the ice is breaking up earlier in the spring. Arctic Ocean ice is decreasing in thickness as well as in area of minimum extent almost annually. Glaciers are retreating in most parts of the world. (Oerlemans 2005). Glacier National Park will soon be that only in name. In the tropics where glaciers occur on high mountains like Kilimanjaro and Kinabalu, all tropical glaciers are retreating at such a rate that all tropical glaciers will be gone in less than 15 years (Thompson et al. 2011).

Another physical change is sea level rise (Church and White 2006). Originally just from the thermal expansion of water, melting of ice on land (e.g. Greenland) is contributing to total ocean water and thus rising sea level. On the Eastern Shore of Maryland, where sea level rise is compounded by natural subsidence of the land, the Blackwater National Wildlife Refuge is on its way to becoming a marine refuge. Just a bit farther south in the Norfolk and Virginia Beach Area of southern Virginia, flooding from sea level rise is increasingly frequent.

Another aspect of the physical impact is the increased frequency of intense weather events such as torrential downpours. Tropical cyclones seem to be increasing in intensity but not in frequency. There is no question about the increased frequency of wildfire in the American west as a consequence of longer, dryer summers, and often less accumulation of winter snow pack.

Biological Impacts of Climate Change

Biological impacts of climate change are pervasive. Certain flowering plant species are blooming earlier at The Royal Botanic Garden at Kew in the United Kingdom. Lilacs are blooming earlier in New England. Long term records are rare but where they exist as at Walden Pond or at Aldo Leopold's "shack" in Wisconsin, they all show consistently earlier phenology (timing of natural cycle events like blooming) (Ellwood et al. 2013). In addition animals are changing their life cycles, e.g. earlier tree swallow nesting and egg laying in the upper Midwest.

An even more important development is that some species are beginning to change where they occur geographically. One of the first so documented was the Edith's Checkerspot Butterfly, one of the two most studied species of butterflies in North America. It has clearly been moving upward in altitude and northward as it seeks its required conditions (Parmesan 1996). Similar patterns have been documented in European butterfly species. Various tree species now occur at higher altitudes on the Amazonian slopes of the Andes.

The Joshua tree has now established itself outside the Joshua Tree National Park. The National Arbor Day Foundation, the purpose of which is to encourage planting trees, has found it necessary to produce a new hardiness zone map (2006) to guide successful planting.

Distributional changes are also occurring in the oceans. Plankton distributions are changing. Fish species distributions are changing. Interestingly distributional changes seem to be more rapid than on land, perhaps reflecting many organisms not fixed to a substrate (although coral reefs are an obvious exception) and the ease of moving in a liquid medium. In the Chesapeake Bay, America's great estuary, eel grass communities so important for seafood productivity are very temperature sensitive so the southern and northern distribution limits have been shifting northward year after year.

These kinds of effects are not being observed just in the temperate or boreal regions; they are also being detected in the tropics. There the effects may be more tightly tied to moisture availability. In the legendary cloud forest of Monteverde in Costa Rica, the primary change is the frequency of cloud formation. Cloud formation is increasing at higher altitudes with more dry days in Monteverde. That can affect the very existence of cloud forest ecosystems which depend on condensation from clouds as the almost exclusive source of water.

Nature is rich in tightly timed co-evolved "coupled" relationships between species. As a consequence of the 10,000 years of climate stability, the timing of some these involve one element which uses day length as the measure of time whereas another may use temperature. As those diverge in timing under climate change (with the day length not changing but temperature changing due to climate change) the consequence is called a decoupling event. A prime example is the snowshoe hare which uses day length to change from winter (white) to summer (brown) pelage (and back) whereas the environment the pelages are designed to match are temperature sensitive. There is little flexibility in the snowshoe hare's system with the consequence that it is frequently occurring in white pelage in a snowless environment, and hence is very vulnerable to predation.

Another example of a decoupling event involves a seabird, the Black Guillemot, which is a colonial nesting bird on the shores of the Arctic Ocean. Its primary food source is the Arctic cod which occurs at the edge of the sea ice. With warmer summers this has meant longer trips from the nesting colony to the edge of the ice for food for its chicks as well as for itself. That has become a longer and longer trip with the consequence that at least one nesting colony has failed.

Expected Future Impacts of Climate Change

The foregoing are essentially minor ripples in the fabric of life. Nonetheless the changes that have already occurred have clearly gone beyond the realm of individual example and anecdote. The impact of climate change on biodiversity is pervasive and statistically robust.

The more important question is this: what can we expect going forward?

For well-known species it is possible to construct a climatic envelope, namely an account of the conditions in which the species currently occurs. Then casting ahead with climate models a future distribution of that envelope can be estimated. For the sugar maple, the required climatic conditions at double pre-industrial levels of CO_2 are likely to be found only in Canada (according to all five of the major climate models) and not in northeastern United States where it is so prominent today.

The impact of climate change on conditions for species is not a matter of temperature alone. Moisture is also important. Those two factors are the most important for terrestrial species and systems. For aquatic ones the two most important are temperature and pH or acidity. All of these are changing.

For freshwater ecosystems, temperature can be very important. Cold water species like trout but also aquatic invertebrates will be seriously challenged.

Species which live at high altitudes are also specially challenged as a group. Moving upslope works as long as there is habitat upslope. An example would be the American pika which lives in isolated colonies in the southern Rockies; the various separate colonies have been moving upwards. There has been a 44% decline of pika in the Great Basin (USGS).

Coastal species will have to cope with sea level rise, and in some instances there will be no way to retreat inland as an uninhabited marsh might have permitted in the past.

Island species are also threatened as a group. Those on low lying islands like the Key Deer will simply have their habitat disappear underwater. Yet even species on more elevated islands are likely to be vulnerable to the loss of suitable climate.

So high altitude species and island species are two classes of more immediately vulnerable species. So too are those with natural histories linked to ice, such as the polar bear.

Looking yet further ahead and assuming a lot more climate change, there are some challenging complications.

While we know that there has been significant advance and retreat of glaciers in Pleistocene North America and Eurasia driven by past climate change, it seems that those changes occurred without any significant loss of biodiversity. Species were basically able to track their required conditions and move across landscapes with them.

The difference today—the first challenge—is that landscapes are heavily modified by human activity essentially creating obstacles to dispersal. Fortunately that is something relatively easily dealt with by active and programmatic restoration of natural connections in landscapes, such as restoring riparian vegetation. Vegetation along watercourses and around waterbodies has large benefits in prevention of soil erosion and improvement in water quality so they are desirable in themselves as well as having biodiversity benefits.

The remaining challenges are tougher. The second challenge is that we know with a significant amount of climate change that entire ecosystems or biological communities do not move as units. Rather the individual species respond, each moving in its own direction and at its own rate. The consequence is that many ecosystems as we know them will disassemble, many species will become extinct, and the survivors will assemble into communities both difficult to imagine or manage as a process.

The fundamental point is with greater climate change the management challenges rise exponentially not linearly—and are best avoided.

A third challenge is that in contrast with the computer models which are gradual and mostly linear, we can anticipate some nonlinear and abrupt climatic change. That certainly has happened climatically in the past when the "global conveyor belt", which distributes heat around the globe through the oceans, shut down.

It is both interesting and disturbing that we are already seeing some abrupt change in ecosystems. A prime example is the massive dieback of trees in the coniferous forests of western North America from southern Alaska to southern Colorado. Warmer winters and longer summers have tipped the balance against the trees and in favor of the native bark beetles, allowing an additional summer beetle generation and much greater overwintering beetle survival.

A yet more striking example is occurring in the oceans where only a brief period of somewhat warmer water affects the basic partnership at the heart of coral reef ecosystems, namely between the coral animal and an alga. The coral animal ejects the alga causing "beaching" events leading to the collapse of diversity and productivity as well as the benefit to local communities. That can be minimized by reducing other stresses on the coral reefs but it can't be eliminated.

What the above are demonstrating is there are real limits to what we can learn from climate models and vegetation models. We can anticipate there will be other relationships between just two or a few species which will be more sensitive than anything those models can reveal. We should anticipate surprises.

The final complication is change at even greater scale, which we can think about as system change. One example would be the vulnerability of the hydrological cycle of the Amazon which generates about half the rainfall in the Amazon basin. That depends not only on moisture entering from the tropical Atlantic but also on the forest itself, the complex surfaces of which lead to major evaporation after a rainfall together with moisture transpired by the trees. It is a complex relationship which recycles the water five to six times before it reaches the Andes.

The Amazon hydrology therefore is vulnerable to climate change, as well as deforestation, and indeed the synergy between the two. It could lead to "Amazon dieback," with a major loss of biodiversity and carbon, as well as degradation of quality of life for people living in the Amazon.

An even greater form of system change was pretty much overlooked until about 2005, namely acidification of the oceans. Basically as the oceans absorb carbon dioxide some of it turns into carbonic acid and affects the carbonate equilibrium. The oceans today are about 0.1 of a pH unit more acid than in pre-industrial times (Acidity and alkalinity are measured by pH but it is on a logarithmic scale so 0.1 is equivalent to 30% more acid).

The acidification has highly significant implications for the myriad species that build their shells or skeletons from calcium carbonate. This has already created problems for oyster beds in the northwest United States and is affecting the tiny sea butterflies or pteropods which serve as the base of the food chain for major fisheries off Alaska or in the North Atlantic.

In sum, as climate change increases we are ever closer to affecting critical and very poorly understood thresholds in the Earth System. So it becomes very important to address the problem while it is still in the manageable range.

Solutions for Climate Change

It is clear that the 2° limit long held out as the target for limiting climate change is actually dangerously high. A world in which this world has occurred would be a world without tropical coral reefs and there would be a lot of other bad ecological change and biodiversity loss.

From a sea level perspective, the last time the planet was 2° warmer the oceans were 4–6 m higher (Kopp et al 2009). No wonder the small island states led the way at the 2016 Paris Conference of the Parties to resurrect a goal of 1.5° (or about 350 ppm).

So what to do? Obviously there is a major energy agenda to move society away from fossil fuels as rapidly as possible and toward various kinds of renewables. Anything that reduces emissions or pulls CO_2 from the atmosphere can contribute toward a better outcome.

There also has to be a major conservation agenda to adapt to climate change, such as improving connectivity in the natural environment so species can actually follow their required conditions. Anything that reduces other stresses on biodiversity and biological systems contributes to a better outcome.

There also needs to be a major effort to reduce CO_2 emissions from deforestation. The annual gross CO_2 emissions from deforestation is about 30% of all emissions, greater than the entire transportation sector.

That is not enough, however, to get us back to and keep us below the 1.5° limit, but the biology of the planet can actually help us achieve that.

A Wild Solution

In the geologic past there were two occasions on which there were extremely high atmospheric CO_2 levels that then reduced to pre-industrial levels by natural processes. The first happened when green plants arrived on land and the second coincided with the arrival of modern flowering plants. So we know the planet can lower the current high CO_2 concentrations through natural processes. But these natural processes in the geologic past occurred over tens of millions of years and we don't have the luxury of waiting tens of millions of years for it to happen.

Unbeknownst to most people there is a significant amount of carbon currently in the atmosphere that comes from centuries of ecosystem destruction and degradation. And if those ecosystems are largely restored it would pull a significant amount of carbon out of the atmosphere before it traps the equivalent radiant heat and causes the consequent climate change. Scientists at Woods Hole are working on a map of ecosystem restoration potential but a rough calculation shows the carbon to be in the range needed.

Forests immediately come to mind, but there are significant amounts of degraded grazing land in the world. Restored grasslands would support better grazing. Agro-ecosystems that accumulate instead of leaking carbon would actually become more fertile, and more recently the potential of restoring coastal wetlands and ecosystems like mangroves has been shown to have significant potential. Soils are the hardest to derive a good carbon figure for in part because soils and their carbon are so patchy in distribution, but they also can have large potential as indicated by the original deep American prairie soils.

Such an effort would be congruent with the Half-Earth proposal by E. O. Wilson and the convergent Nature Needs Half proposal both of which are driven by biodiversity conservation imperatives. All those biodiversity conservation efforts have carbon sequestration benefits as well.

In addition to bringing climate change down to a level better for the biology of the planet and for human well being, there are two additional benefits to the restoration agenda. One is that it will help people understand that the planet does not work solely as a physical system, but rather it works as a linked biological and physical system and is indeed a living planet. The second would be the recognition that the restoration agenda suddenly turns the climate challenge from "what can I possibly do about such a big problem?", to one where individual action like planting a tree—always a desirable action but now recognized as a benefit to the planetary system—will empower individuals much as victory gardens did in World War II.

Conclusions

The growing impacts of climate change on biodiversity present a compelling case for limiting global warming to 1.5° greater than pre-industrial levels. A wild solution—restoration of the extensive degraded and destroyed ecosystems over past centuries—can make a major contribution to achieving such a goal.

References

Arrhenius S (1896) On the influence of carbonic acid in the air upon the temperature of the ground. Philos Mag Ser 5, 41(251):237–275

Church JA, White NJ (2006) A 20th century acceleration in global sea-level rise. Geophys Res Lett 33(L01602):1–4

Ellwood ER, Temple SA, Primack RB, Bradley NL, Davis CC (2013) Record-breaking early flowering in the Eastern United States. PLOS. https://doi.org/10.1371/journal.pone.0053788

Kopp RE, Simons FJ, Mitrovica JX, Maloof AC, Oppenheimer M (2009) Probabilistic assessment of sea level during the last Interglacial Stage. Nature 462:863–867

Oerlemans J (2005) Extracting a climate signal from 169 glacier records. Science 308:675–677

Parmesan C (1996) Climate and species' range. Nature 382:765–766

Thompson LG, Moseley-Thompson E, Davis ME, Brecher HH (2011) Tropical glaciers, recorders and indicators of climate change, are disappearing globally. Ann Glaciol 52:23–34

Printed in the United States
By Bookmasters